A YEAR ON THE RIVER

FIONA SIMS

ADLARD COLES
Bloomsbury Publishing Plc
50 Bedford Square, London, WC1B 3DP, UK
Bloomsbury Publishing Ireland Limited,
29 Earlsfort Terrace, Dublin 2, D02 AY28, Ireland

BLOOMSBURY, ADLARD COLES and the Adlard Coles logo are trademarks of
Bloomsbury Publishing Plc

First published in Great Britain 2026

Copyright © Fiona Sims, 2026
Illustrations © Lilly Louise Allen, 2026
Photographs © Gary Latham and Fiona Sims, 2026
The photo on page 9 is reproduced with permission from Robbie Cumming
The photo on page 42 is reproduced with permission from @rokabringsflowers

Fiona Sims has asserted her right under the Copyright, Designs and Patents Act, 1988,
to be identified as Author of this work

For legal purposes the Acknowledgements on p.205
constitute an extension of this copyright page

All rights reserved. No part of this publication may be: i) reproduced or transmitted in
any form, electronic or mechanical, including photocopying, recording or by means of
any information storage or retrieval system without prior permission in writing from the
publishers; or ii) used or reproduced in any way for the training, development or operation
of artificial intelligence (AI) technologies, including generative AI technologies. The rights
holders expressly reserve this publication from the text and data mining exception as per
Article 4(3) of the Digital Single Market Directive (EU) 2019/790

Bloomsbury Publishing Plc does not have any control over, or responsibility for, any third-
party websites referred to or in this book. All internet addresses given in this book were
correct at the time of going to press. The author and publisher regret any inconvenience
caused if addresses have changed or sites have ceased to exist, but can accept no
responsibility for any such changes

A catalogue record for this book is available from the British Library

Library of Congress Cataloguing-in-Publication data has been applied for

ISBN: PB: 978-1-3994-1975-8; ePub: 978-1-3994-1974-1; ePDF: 978-1-3994-1973-4

2 4 6 8 10 9 7 5 3 1

Designed by Nicola at Big Orange Door
Typeset in Archer and Poppins
Printed and bound in China by RR Donnelley Asia Printing Solutions Ltd

To find out more about our authors and books visit www.bloomsbury.com
and sign up for our newsletters
For product safety related questions contact productsafety@bloomsbury.com

A YEAR ON THE RIVER

A BOATER'S GUIDE TO GROWING, FORAGING, STORING AND COOKING

FIONA SIMS

FOREWORD BY ROBBIE CUMMING

ADLARD COLES

LONDON • OXFORD • NEW YORK • NEW DELHI • SYDNEY

CONTENTS

Foreword by Robbie Cumming 8

Introduction 10

1 BOATING LIFE

The galley 15
Kit 16
Store cupboard 20
Eating local 30
Seasonal eating 33
Foraging 34
River fishing 37
Growing your own on board 40
For the love of honey 45
Pickling and preserving 48
British drinks 59
River rescue 65
Sustainable boating 70
Boat rental 74
Narrowboat heritage 80
Boat barbecues 88
Boat picnics 94

2 SPRING

What's in season? 98
Spring foraging 98

RECIPES

BREAKFAST
Porridge with rhubarb and orange compote
  and toasted almonds 100
Naan bread breakfast pizza 101

QUICK & LIGHT
Wild garlic pesto 103
Asparagus and new potato frittata 105

BIGGER PLATES
Chicken, spinach and lemon orzotto 107
Prawn and chickpea stew with leeks and lemon 108

SWEET TREATS
Alphonso mango with coconut and mascarpone 111
Sea salt chocolate brioche toasties 112
Peanut butter biscuits 112

TIPPLE TIME

What to drink 116
Three great spring cocktails 118

BEER SNACK
Prosciutto, crisps, Guindillas peppers 122

3 SUMMER

What's in season? 126
Summer foraging 126

RECIPES

BREAKFAST
Boat granola 129
Banana pancakes with wild blackberries 130

QUICK & LIGHT	Hot smoked salmon salad with watercress and fennel	132
BIGGER PLATES	Puy lentils with roasted sweet peppers, tomatoes and feta	133
	Panzanella	134
	Roasted chalk stream trout with fennel, herbs and citrus	136
	Rosemary, garlic and honey pork chops	138
SWEET TREATS	Big scone, strawberries, and clotted cream	139
	Grilled peaches with honey mascarpone and rosemary	140

TIPPLE TIME

	What to drink	143
	Three great summer cocktails	144
BEER SNACK	Crudités with herbed crème fraîche	148

4 AUTUMN

What's in season? 152
Autumn foraging 152

RECIPES

BREAKFAST	Strapatsada	154
	Avocado and dukkah toast	156
QUICK & LIGHT	Mushroom kimchi rarebit	156
	Tomato tapenade tart	159
BIGGER PLATES	Greek spinach rice and grilled lamb chops	160
	Minced venison and oyster mushrooms	163
SWEET TREATS	Basque cheesecake	164
	Honey-roasted figs	166
	Plum and coconut cake	167

TIPPLE TIME

What to drink **169**

Three great autumn cocktails **170**

(BEER SNACK) Black bean and corn quesadillas **174**

5 WINTER

What's in season? **178**

Winter foraging **178**

RECIPES

(BREAKFAST) Tofu, kimchi and kale scramble **181**

Chickpeas, chorizo and eggs **182**

(QUICK & LIGHT) Leek and butter bean soup **183**

Fi B's fishy rice **185**

Broccoli and anchovy toasts **187**

(BIGGER PLATES) Truffade **188**

Korean beef stew **190**

(SWEET TREATS) Blood orange and poppy seed muffins **194**

Chocolate, pistachio and apricot tiffin **195**

TIPPLE TIME

What to drink **197**

Three great winter cocktails **199**

(BEER SNACK) Montgomery Cheddar and rosemary popcorn **202**

British cheese **204**

Acknowledgements **205**

Index **206**

FOREWORD

BY ROBBIE CUMMING, NARROWBOATER AND TV PRESENTER

Cooking on a canal boat – much like living on one – isn't easy. Space in general can be extremely limited. Not being able to fit much in the oven, the lack of mod cons, and the inability to buy in bulk to save money as cupboard space is at a premium.

Limitations with the kit you have or the ingredients you can store can, as Fiona shows in this book, be a springboard for creative recipe ideas. In a galley you can end up creating something unique that you'd never have thought of in a kitchen.

On board my narrowboat, *Naughty Lass*, I like to keep things simple. I have three gas hobs, a gas oven and a small 12v fridge freezer. Pots and pans, ladles, tongs, knives – whatever will hang up – I keep up out of the way to make room for jars, cans, plates and trays. I don't have a microwave or a dishwasher or a food processor – but I don't miss either of these things. You get used to making do with what you've got.

I love being so close to nature as I cook. It teaches you to respect what you put down the drain. I often cut out the drain and go straight to the source, leaning out of my hatch to drain pasta, or throw out small veg peelings to the ducks, swans and fish who to be honest would usually prefer a somewhat tastier soda bread (see pages 20–1 for Fiona's recipe).

I'm normally overwhelmed by recipes. Or put off by the foodie snobbery and too-posh sounding ingredients. This book, however, is a personal take on food from an author who appreciates the limitations of cooking in a small space, not to mention the joy it can bring when shared with friends and family.

For those that are adventurous enough, dive headfirst into this book and dredge up as many mouth-watering recipes as you can. Whether you're a landlubber or a ditch-dweller, I'm sure you'll find something here to feed the mind, soul and – most importantly – belly.

INTRODUCTION

It was love at first sight when I saw *Ellenene*. Moored on a peaceful stretch of the river near Oundle in Northamptonshire, she promised a weekend of much sought-after calm. Swans glided by, unfazed by a coxed four slicing through the water on their morning training session, while a kingfisher darted by in a streak of metallic blue.

Sporting a new burgundy and grey livery, the 1995-built narrowboat offered up its charms in an instant, the kettle whistling its welcome as enamel mugs of tea beckoned, cosy sheepskin-draped seating arranged around a crackling wood-burning stove. This was boating, but not as I knew it.

If you're familiar with my earlier publications, *The Boat Cookbook* and *The Boat Drinks Book*, then you'll know that I'm more used to the open seas – not the gentle flow and pace of our inland waterways, which couldn't be more different. Instead of living on your wits as you tack and jibe, your focus entirely on the wind and tide, the river demands a more inward-looking perspective, inducing tranquillity and a sense of peace – and it was just what we needed at time of turmoil (yes, Covid, I'm looking at you).

My brother-in-law, Kim, and sister-in-law, Liz, had bought the narrowboat in 2018, as an escape from the hustle and bustle of city life. When lockdown partially lifted, we accepted their invitation to stay on board for a weekend and it made a lasting impression. The plan was that they would spend the day showing us the ropes, so to speak, then leave us to enjoy river life – which we did, and then some. Now they can't keep us away.

It was during these subsequent visits that Kim and Liz asked me, why not write a food and drink book for them, and other river boaters? So here it is, including tips on foraging along the riverbank and its surrounds, pickling and preserving, growing your own on board, celebrating the seasons, the art of buying local, and even how to live more sustainably on the water.

The 40 or so recipes in the book are mostly mine, with the rest extracted from a handful of top chefs who have generously shared their imaginative recipes that will set your galley cooking apart. Most of the dishes can be completed in 20-30 minutes tops, and most are enough for four hungry boaters. Virtually all the ingredients are available at the supermarket or local shops, and where they aren't, I've included online contact details.

I've divided the recipe part of the book into the seasons, so that it's more useful for boaters, hopefully making everyday life on board easier. The rest doesn't necessarily follow a logical path, rather it meanders through subjects relating to river life – not unlike a typical day on the water – but it's all there listed in the contents if you need a navigation tool.

It turns out that I'm not the only one to be love-struck by our inland waterways. Huge numbers of us are taking to the UK's rivers and canals, which are as busy as they ever were, if not busier. Not least, a multitude of television crews are filming on our waterways (with Robbie Cumming's BBC show *Canal Boat Diaries* the best of the bunch), making programmes lapped up by surprising numbers as we sit at home enjoying the lifestyle vicariously, absorbing the more zen moments, which instil a sense of sought-after calm.

Meanwhile, the restoration of our inland waterways is continuing at pace, with over 100 groups of volunteers working on reopening old canals and even planning new ones. And people can't get enough, aspiring to live on them or near them, or just hanging out on them, especially in cities. The golden age of our canals wasn't the 1800s, it's now.

Meanwhile a growing band of campaigners are fighting to keep our rivers clean with various clever initiatives now in place. There's still a way to go, but the message is loud and clear – we want our rivers back, we want the wildlife back, and we need these spaces to help clear our heads. Add to all that a cracking plate of something delicious and locally grown to wash it all down and you have achieved boating nirvana. That's my take on it.

Boating Life

Living on the river is a unique, relaxing, rewarding experience, providing a slower pace of life and a chance to connect with nature. But you need to be organised. You can't always just pop to the shops if you need something, so it pays to think ahead and keep your galley stocked. And then there is the free food all around you – the riverbank, hedgerows and woodlands are teaming with fresh, nutritious produce. Let me be your guide.

THE GALLEY

These days a galley on your average narrowboat or cruiser is pretty well equipped. Yes, space will be tight but boat builders are ingenious space savers, with every nook and cranny utilised, gadgets and storage cleverly concealed. And if you own a newer model boat, then you've probably got an even better galley on board than my kitchen at home.

Kim and Liz's galley on *Ellenene* has all the essentials, with an oven that's just a bit smaller than your average household oven, plus a three-ring hob. They might not own a boat fridge yet (unlike their neighbour, Anna, who lives on board full-time and runs a 12-volt fridge powered by solar energy), but they make do with a cool box and store plenty of dried and tinned food on board for longer stays, supplementing with fresh ingredients bought from the local shops.

As space is at a premium on board, and everything has its place, the kit you select should be chosen carefully. Pots and pans should stack easily, and dishes should be both oven and table friendly. There are lots of space-saving options around, so take advantage of the latest clever kitchenware.

It's a good idea to have a master list of everything you keep on board, then every time you finish something, cross it off so you remember to replace it the next time you shop. Buy as you go, keeping only basic items on board such as flour, sugar and cereals, plus a few store cupboard staples for when you're caught short. Have a store cupboard supper or two stashed away for when you might arrive empty-handed late in the day. I'm not suggesting that you stock up on all of the things that I've listed overleaf, it's just to give you a few ideas and hopefully discover something new – hello, za'atar.

KIT

AIR FRYER I'm the first to dump a new bit of kit after the novelty and usefulness wears off, but this one is a keeper. Fluffy roast potatoes and roasted vegetables in 15 minutes; crispy chickpeas to add crunch to salads; salmon cooked in six minutes. I could go on. You might be thinking, this is just another gadget that sucks up energy, but air fryers use less energy than conventional ovens, cooking things in half the time. They're a healthier way to cook too as they use hot air and less cooking oil, or sometimes none at all. And some things actually taste better cooked in an air fryer – chips, toasted sandwiches and roast chicken to name three. For cooking on board for those with generators or regularly hooked up to electricity, then try the dinky, cute-looking Wonder Oven from Our Place, a six-in-one marvel that steams, toasts, bakes, roasts, reheats and air frys.

BAKING TIN One that fits your oven size exactly. Non-stick helps. Roasting bags are another useful addition.

BARBECUE Barbecues and boating go hand in hand, especially on a warm summer's evening. If you're cooking for two you only need a small amount of heat and a small, light barbecue is just fine. Forget disposable ones as the quality of the charcoal is poor, the smoke toxic, your cooking window brief and the flavour disappointing. The classic kettle has changed little since the 1950s, while more modern shapes abound. Fill with good quality lumpwood charcoal.

CHOPPING BOARDS Choose flexible, colour-coded synthetic boards to avoid any cross contamination when chopping raw meat and fish.

COCKTAIL SHAKER Don't laugh. It's essential if you love a cocktail on board. Unbreakable, too, if you choose a stainless steel one, although you can improvise and use a Thermos.

COLANDER Stainless steel is the most practical, but vintage enamel the prettiest.

CONTAINERS Decant your coffee, tea and sugar into durable containers, steering clear of glass. Re-use supermarket jars of herbs and spices, et al.

CUTLERY The usual set of knives, forks and spoons, plus a wooden spoon, fish slice, scissors, bottle and can openers, vegetable peeler, small balloon whisk, grater (microplane is the best), and skewers.

KNIVES Buy good ones, keep them sharp, and store them safely, it will make life on board easier and cut down on preparation time. A 13cm cook's knife will work for most tasks.

KETTLE Choose one you can fill through the spout for ease.

MANDOLIN Sounds a bit fancy but you can buy a cheap version (Oxo Good Grips does a plastic one for around £10), and it will make salads more exciting.

MEASURING CUPS/JUG So many recipes now come with both the metric and (American) cup system, especially online. Failing that, use a standard mug – it holds 225 ml of liquid or enough rice for two to three people. Or just guess, which I do frequently with varying accuracy.

MIXING BOWLS You will need a couple. Best is a set of stainless-steel mixing bowls that will fit inside the pressure cooker (see below) for easy storage.

OVEN THERMOMETER Because some galley stoves are a tad tricky to regulate and thermostats are often unreliable.

POTS & PANS One deep-sided sauté pan with a lid, ideally around 28cm in diameter, plus a smaller saucepan, both preferably with removable handles for easy storage. Don't stint on your pans. Heat distribution is the name of the game here. I'm loving the Our Place '8-in-1' Perfect Pot on board for its lightness, durability, multi-function prowess and, frankly, good looks (fromourplace.co.uk). You'll also need a (toxin-free) non-stick frying pan – far easier on the washing up. Little tip: measure the top of your hob first and check that the pans will be able to sit together in most combinations.

PRESSURE COOKER It's your fastest route to a sustaining meal on board. It transforms dried beans into soups and stews in just 20 minutes and makes quick work of connective tissue – flavours tend to be better too. It saves on fuel and will survive an errant wake. Invest in a good, old-school gas hob-friendly one, made from stainless steel. The pressure cooker also doubles up as a pasta pan, or indeed anytime you need to use a deep saucepan – just find another lid for it. There's a bit of trial and error, which involves working through watery gravies and mushy veg, but there are some good books around to guide you, such as *Everyday Pressure Cooking* by Catherine Phipps (Quadrille).

RAMEKINS Sure, you can just use one big dish, but dividing the recipes up into individual portion sizes not only cuts down the cooking time, but they double up as serving dishes, so less washing up. Use metal ones on board.

SPLATTER GUARD Because the last thing you want in a confined space is smelly grease splats.

TABLEWARE Hard-wearing enamelware is the perfect boater's crockery. I love those from Falcon Enamelware, which come in an array of vibrant colours (falconenamelware.com). I always look out for pre-loved enamelware mugs in charity and vintage shops. Make sure you include deep-dish bowls for porridge, soups and noodles, and choose stackable drinking tumblers to save on space.

THERMOS For keeping those hot dogs at the ready when you're on the move, for warm soup when you need it, for shaking a cocktail, and for a ready supply of hot toddies for a cosy beverage under the stars.

STORE CUPBOARD

You don't need all of these (and won't necessarily have the space), of course, it's just a guide to give you some ideas. You will need some of these ingredients for the following recipes, but you can pick them up as and when needed.

As far as the fresh basics go, I take on board a combination of these ingredients: sourdough bread (because it tastes good and lasts so long), eggs, milk, butter, Parmesan, feta, lemon, garlic bulb, fresh herbs, chorizo, onion.

BAKING INGREDIENTS   Flour (plain, wholemeal or both), baking powder, caster sugar.

OATY SODA BREAD

The perfect boat loaf when you can't get to the shops, Irish soda bread is quick and easy to make and it doesn't contain yeast so doesn't have to rise. And if it's good enough for Robbie Cumming – the star of BBC's *Canal Boat Diaries*, who baked it once while on board his narrowboat, *Naughty Lass* – then it's good enough for me. I add porridge oats (plus an extra sprinkle on top) for texture. Instead of traditional buttermilk, I use an equal mix of readily available yoghurt and milk. Soda bread is best eaten on the day it's made, preferably with lashings of butter.

250g plain white flour
250g plain wholemeal flour
100g porridge oats

1 level tsp bicarbonate of soda
1 level tsp salt
25g butter, cut into small dice
250 ml milk
250 ml natural yoghurt

Preheat the oven to 200°/fan 180°/gas 6. Mix dry ingredients together in a large bowl then rub in the butter. Combine milk and yoghurt, pour into the dry ingredients, then mix in quickly with a round-bladed knife. Bring the soft dough together gently with your fingertips. Shape into a flat, round 20 cm loaf and place onto a flour-dusted baking tray. Score a deep cross on the top if you like (it lets the fairies out). Bake for 30 minutes or until the bottom of the loaf sounds hollow when tapped. Transfer to a wire rack and leave to cool.

CEREALS Jumbo porridge oats, homemade granola (see page 129).

GRAINS & PULSES Pearled spelt (it cooks faster than barley), basmati rice, quinoa, couscous, Puy lentils (pre-packed, ready-cooked for ease), brown lentils (tinned), red lentils, quick cook polenta.

HERBS, SPICES, SALT & PEPPER Smoked paprika (it gives a meaty depth to soups and stews). Ras el hanout Moroccan spice mix (try rubbing it on lamb chops). Za'atar (this wonderfully complex Middle Eastern spice mix, which varies from region to region and brand to brand, will transform eggs, pimp up salads, and add a new dimension to soups). Sumac, curry powder, garam masala, cinnamon, thyme, rosemary, ginger, crushed chillies, cumin, coriander, turmeric, fennel seeds, nutmeg, bay leaves, bouquet garni, saffron, black pepper, sea salt (when I list salt in the recipes, I mean sea salt rather than table salt, unless you are salting water for pasta – Maldon is particularly good), flavoured salts.

DUKKAH

Dukkah is your store cupboard secret weapon. I sprinkle it on pretty much anything, from roasted veg, to crunchy green salad, soft-boiled eggs, and hummus. I especially like it for breakfast. What is it? It's a Middle Eastern nut, seed and spice blend that began life in ancient Egypt. The word 'dukkah' comes from the Arabic verb 'to pound' – a nod to the traditional method of making it by grinding it up in a pestle and mortar, which you'll need for this recipe (but if you don't have one then use a small bowl and the end of your rolling pin). You don't need much, just a tablespoon sprinkled over whatever you fancy. It's healthy too – a simple way to add protein, fat and flavour to a meal. You can use other nuts too of course. Cashews and walnuts are also good, but I like this combination best. Do what they do in the Middle East and first dunk a piece of bread into a bowl of extra virgin olive oil, then dip into the dukkah for a quick, delicious pre-dinner nibble.

MAKES A SMALL JAR

75g shelled unsalted pistachios
75g whole skin-on almonds
1 tbsp coriander seeds
1 tbsp cumin seeds
4 tbsp sesame seeds
Handful of fresh mint leaves, chopped (optional)
1 tsp chilli flakes
1 tsp Maldon salt

Preheat the oven to 200°/fan 180°/gas 6. Scatter the nuts on a baking tray and roast for five minutes, or until they are just starting to turn golden. Cool, then chop roughly. In a non-stick frying pan over a medium heat, warm the coriander and cumin until they begin to release their aroma. Transfer to the mortar, and bash with the pestle until roughly crushed. In the same pan, lightly toast the sesame seeds. Combine the nuts, seeds and spices in a bowl and stir in the fresh mint, if using, the chilli flakes and crumble in the salt. The dukkah will keep for 2-3 weeks in a screw-topped jar.

NUTS, SEEDS & DRIED FRUIT Almonds, walnuts, pine nuts, cashews, pistachios, peanuts, mixed seed sprinkle, dried apricots, prunes and sultanas.

OILS & VINEGARS Olive or rapeseed oil (I'm talking extra virgin. I use them in pretty much everything, and increasingly choose rapeseed as it's British, cheaper, just as healthy and tastes almost as good as anything from Tuscany.) Sunflower oil for a lighter touch.

Balsamic (love the Belazu brand, widely available) and white wine vinegar.

> **THE PERFECT VINAIGRETTE**
> In a screw top jar combine 2 tsp Dijon mustard, 2 tbsp of white wine vinegar, 6 tbsp olive or rapeseed oil, 1 tsp honey and a pinch of salt. Shake until emulsified. Store in the fridge or cool box until needed.

ODDS & SODS Dried porcini mushrooms to tart up bog standard mushrooms and add depth to soups and stews, sun-dried tomatoes, tofu (it comes in a small handy box with a long shelf life and has an ingenious ability to sop up flavours and keep you healthy), maple syrup, 70% dark chocolate.

OLIVES & PICKLES Go for superior olives, preferably those packed in extra virgin olive oil, jalapeño peppers (to transform hot dogs, pizza, and eggs), capers.

PRESERVES & SPREADS Marmalade, honey, Marmite, peanut butter, Nutella, salted caramel sauce. The last two are seriously good stirred into yoghurt for an instant dessert.

RICE, PASTA & NOODLES Basmati rice, risotto rice, angel's hair pasta, macaroni, orzo, penne, spaghetti, egg noodles, rice noodles, udon noodles.

SAUCES & PASTES Horseradish, hot sauce, ketchup, mayonnaise, mustard (Dijon and English), passata, pesto, tamari soy sauce, salsa, Tabasco (for Bloody Marys), tomato sauce, Worcestershire, laksa paste, miso, tapenade, tomato paste.

STOCK Marigold vegetable bouillon (which I use anytime stock is required).

TEA & COFFEE Tea bags (including herbal), ground and instant coffee, not forgetting cocoa powder, powdered milk and UHT milk.

TINS & PACKETS Anchovies, baked beans, beef consommé, chickpeas, chopped tomatoes, tomato soup (Heinz, of course), coconut milk, corned beef, tinned pâté, tuna, sardine fillets, white crabmeat, artichoke hearts, prunes, chickpeas, butter beans, cannellini beans, borlotti

beans, haricot beans, evaporated milk (try adding to porridge in place of milk for extra creaminess), vacuum-packed wholemeal pitta bread, vacuum-packed wholemeal flour tortillas (see page 174), long-life fruit juice, popping corn (see page 202).

FANCY TINS

Is there anything more enticing than a cupboard full of shiny, brightly coloured tins to spark culinary joy? The French have got it sussed, with their ready-prepared, excellent tasting cassoulets, pâtés, tinned duck confits – the latter fabulous served with duck fat cooked potatoes for a smart, cheat Sunday roast (you can buy them online from thegoodfoodnetwork.com). For posh pâté try epiceriecorner.co.uk.

Superior tinned fish has become a bit of a thing. The UK's most famous Spanish chef, José Pizarro, even makes a feature of tinned fish on his menus, selling tinned anchovies for an eyebrow-raising £38 at his Bermondsey restaurant Lolo. "People don't understand tinned fish in the UK," says Pizarro. "Take our anchovies, for example. The Reserva Catalina Cantabrian anchovies are something special and rarely seen. Matured in salt for two years, they have a unique colour, aroma and flavour," he enthuses. You can buy the anchovies from his online shop for £22 for 10-12 (large) fillets. And yes, they are that good.

Also check out online the Tinned Fish Market, where all of its fish is sourced from small family businesses in Spain, France, and Portugal, and boast

beautiful tins. I always buy my tuna from Spanish specialist Ortiz; it is line-caught and canned on the day of the catch (from brindisa.com). Sardines on toast are my weekly go-to. Try west country chef Mitch Tonks' brand Rockfish, or invest in top-tier tins of hand-filleted heritage sardines from thefishsociety.co.uk – sustainably sourced in Andalusia from independent fishermen, beautifully wrapped, and priced accordingly. Or if you're feeling adventurous, try Galician speciality *chipirones en su tinta* – squid served in its own ink, delicious served warm with crusty bread (from souschef.co.uk).

Once you've tried Brindisa's Navarrico chickpeas (and its large *judión* butter beans come to that, both from Ocado), there's no going back – they are the Rolls-Royce of chickpeas. I also love organic brand Biona's range of tinned beans and pulses.

Another must-have in my store cupboard is jars of roasted red peppers, which I whizz up with walnuts, crushed garlic, and a smidge of smoked hot paprika for an easy, flavour-packed dip (in place of a food processor on board try Oxo Good Grips chopper). And I can't resist a pretty tin of plum tomatoes, such as those from Mutti (widely available), re-using the tins for growing basil. Cooks & Co, which supply delis and specialist food shops around the country, are a great source of tinned vegetables, such as peeled artichoke hearts in brine, while Fragata does a good stuffed olive.

RAINY DAY SWEET TREATS

You're stuck on board in the rain and want to nibble something sweet and delicious? Order in a selection of the following:

Who doesn't like a rum baba? A boozy number from Petrone is available in a jar from campaniawines.co.uk, perfect served with a squirt of creamy oat, a great long-life stand-in for single cream.

Or how about pineapple with spiced rum? Opies pineapple with Luxardo (from Ocado) is a decadent dessert. Still on the fruity high kicks, try Biona's tinned pear halves, which are wickedly good on pancakes with a spoonful of Nutella. And M&S peach slices in grape juice are always a treat-worthy stand-by. Fancy a fig? Then keep a stash of Cooks & Co tinned whole green figs in your galley cupboard for pimping up your morning breakfast granola bowl, or wrap them with salty prosciutto for an instant, classy, savoury snack.

Annually, I clock up an embarrassing number of mail-ordered boxes of Harris & James individually wrapped, indulgently thick, dark chocolate-covered shortbread (10 per box), sent from its Suffolk base.

And when I'm in Soho, London, there is always a pilgrimage to William Curley Chocolatier to stock up on his dark rochers – crispy, nutty praline clusters coasted in a thick layer of Amedei Toscano dark chocolate – which see me through the cold winter nights.

Ditto Meg's Cottage crumbly Honey Fudge, just the right balance of buttery creaminess and sweet honeyed notes (megscottagefudge.co.uk). Or for something a tad more exotic, try Madam Chang's Original Pandan Kaya (madamchangs.co.uk), a buttery, rich coconut curd that is divine on toast.

EATING LOCAL

Is there anything more satisfying than eating something that you've grown yourself? A quick shake to dust off the soil, then onto the plate with flavours at full throttle. But not all of us have the luxury of a kitchen garden or allotment, so instead we seek out locally-grown produce at the shops.

The definition of local varies somewhat but it generally means food grown in your own country, or county, and for those passionate about it then within a restricted radius, which even has its own moniker, hyper-local.

And hats off to pubs and restaurants that have made it their mission to keep the food miles to a minimum – such as The Pig hotel chain (with 10 properties at the last count, soon to be 11), which continues to pride itself on its 25-mile menu and lists many of its suppliers for all to see. Indeed, the farm-to-table ethos in the hospitality industry is still going strong; the movement originated in California in the 1960s and 70s as a backlash to the processed food of that era.

Attitudes have shifted somewhat since then. Yes, the further food must travel, the more fossil fuels it uses and greenhouse gasses it emits (especially when food is transported by plane), but the impact the growing methods have on climate change is a bigger cause for concern, say the scientists, who tell us that most greenhouse gas emissions – around 61% – come from the production process itself.

In many crops, for example, it's the huge volume of fertilisers and pesticides needed to allow them to grow on a large scale. And we all know about cows and their emission-heavy feeds (and methane-filled burps) contributing to global warming. There's an easy solution – eat less beef.

On a boat, you are reliant on what's available in the villages, towns, and cities you might pass through. That will include, hopefully, a farm shop or two. Farm shops have expanded significantly in recent years, with an estimated 1,581 farm retailers across the country (according to the latest figures from the Farm Retail Association), as consumers seek out quality, locally-grown produce in the community. This is a far more soul-enriching affair in an age where supermarkets have become purely transactional experiences and where specialist counters are thin on the ground.

If you're lucky, you'll find an honesty box. I love everything about the honesty box – it fills me with joy for humankind. I always keep some loose change on hand for exactly that purpose. It's not just about the produce, which is always good, from fresh eggs to vegetables, chutneys to jams, it's the aesthetic, and obviously the huge trust involved in the enterprise. Buying food doesn't get any better than this.

SEASONAL EATING

In this world of faddy diets, it's taken us a long time to remember the best diet of all – seasonal eating. We have been doing it since the dawn of humans, but we lost our way, seduced by the idea that we can eat whatever we want, whenever we want it, with technology and transport making the seasons virtually redundant.

What is seasonal eating, exactly? It's eating food shortly after it's harvested in the local area.

Why is seasonal eating so important? It's better for the environment, as growing and transporting food to make it always available uses lots of energy, which creates damaging CO_2. It boosts the local economy, as choosing seasonal produce over imported produce supports British farmers and local businesses. It tastes better and is nutritionally superior, as it's freshly picked at the point of ripeness and lands on our plates in the minimum amount of time, which helps to optimise the concentration of certain micronutrients and phytonutrients.

Okay, so it's pricier at an organic farmers' market, but seasonal food is by definition more abundantly available, which generally means it costs less. You just need to know what to look out for and when.

What can we do to make sure we are buying our food more seasonally? Read labels, ask questions, do research, and decide that caring for the environment should be a top priority when you buy food. For example, you can minimise buying foods that have been air-freighted, and those that may have contributed to deforestation.

I'm no saint, I admit. While most of what I buy seasonally is British, there are certain seasonal items I buy – and look forward to – that are from overseas, such as Alphonso mangoes and blood oranges. And I buy some everyday produce that clearly can't be grown in the UK, from citrus fruits to bananas. But you can still do your bit by seeking out Fairtrade fruit, for example.

In the UK, we currently produce about 50% of our food, and much less than that when it comes to fruit. So, if we all want to live more sustainably, it follows that we need to dramatically increase the amount of food we grow in this country.

Growing your own is a great place to start, and if that doesn't rock your boat (excuse the pun), then sign up to a local fruit and veg box scheme or seek out local produce at your nearest farm shop or independent shops. See my guide to what's in season at the beginning of each season chapter.

FORAGING

Good, wholesome, wild food literally surrounds us. The countryside – and even towns – contain virtually everything we need for a sustainable, balanced, healthy diet. And it's free for the taking, growing in hedgerows and on pathways, in woodlands and on pastures. Unsurprisingly, foraging has boomed in recent years, kickstarted by top chefs serving foraged food on menus up and down the country.

Chefs love pennywort, a juicy, miniature lily pad-like leaf, which adds an interesting visual element to their finished dishes. So, too, bladder campion, which tastes like freshly picked peas, and works well in salads. Through chefs, I've also discovered a rather pungent little leaf called alexanders (you can use the flowers and roots too), and the wonderfully named, spinach-like Good King Henry, which crops all year round and which tastes rather good with a few drops of oyster sauce.

In fact, there are more than 160 edible wild plants growing in the British Isles. Okay, so many of these plants are strictly survivalist territory. By that I mean fiddly to prepare and only vaguely palatable after hours wandering ravenous in the wilderness – I'm thinking about you, bitter-tasting greater plantain, with your large waxy, ribbed leaves that need shredding, soaking, boiling, and soaking again. But there are plenty of wild foods that are easy to prepare and taste delicious. I've highlighted five of my seasonal favourites at the beginning of each chapter.

Once you start looking, you can't stop seeing them, whether you're out walking or gliding along the water. Foraging is good for you too – and I'm not just talking about the valuable nutrients you get from these ingredients, or the exercise required to find them. Being out in nature is good for our mental health. Foraging makes us slow down as we tune into what is around us. The slower we go, the more we see, and the more we breathe in all that healthy fresh air. Foraging is mindfulness at its soothing best.

And forget paying a premium for superfoods in fancy food shops – they're all around us. Two of the most obvious examples are nettles and dandelions, the former a rich source of vitamin C, among other nutrients, while the latter is credited with a whole range of health benefits, not least an impressive tally of antioxidants. Not sounding so weedy now, are they?

Eating foraged foods also connects us more closely with the seasons, as we learn to appreciate and value the plants that appear at certain times of the year. There is even a seasonal colour code of sorts. In spring, it's the lush greens of edible leaves. In summer the jewel-like colours of raspberries, strawberries, and rose petals. In the autumn, think browns and purples, such as mushrooms, nuts, bullaces (a bit like damsons), and blackberries.

Wild food is also more sustainable than any shop-bought food as it's plucked right on your doorstep, grown without pesticides, and there's not a plastic wrapper in sight.

The National Trust and others do take a dim view of those who walk away with armfuls of foraged foods to sell commercially, reminding us of the harm that intensive foraging does to our fragile ecosystems. But all hail those who focus attentions on abundant invasive species, such as Tommy Banks. The top Yorkshire chef has found a delicious use for the garden scourge, Japanese knotweed, using it to make jam to glaze his homemade chipolatas, and with the added benefit that the roots provide a hit of the antioxidant resveratrol.

That all said and done, foraging for wild food comes with a caveat – you need to know what you're doing. For starters, not all parts of the plant are necessarily edible, and some plants are toxic for one person and not for another. At first glance, some leaves look similar, like foxglove (highly toxic), and comfrey (edible), so if you're in any doubt at all, leave it well alone. In short, do your research, whether that's attending foraging courses, or reading up extensively on the subject.

A good place to look is eatweeds.co.uk, which runs foraging courses countrywide and is one of the leading wild food sites. Ditto wildfooduk.com, which travels around the country holding courses from Kent to Edinburgh. To read? My now rather battered copy of *Wild Food: A Complete Guide for Foragers* by Roger Phillips (Macmillan) has been invaluable, while John Wright's *The Forager's Calendar: A Seasonal Guide to Nature's Wild Harvests* (Profile Books) is well regarded by those in the know.

HOW TO FORAGE – THE BASICS

- Only pick what you can identify with certainty.
- You can pick anything off public land for your own consumption, but you mustn't dig up any roots.
- Get permission from the landowner to uproot anything, though most wild fruit, foliage, flowers and fungi (see my note below) are legally fair game on almost any land.
- Gather only plants that are growing in profusion, picking a little bit everywhere you go rather than taking handfuls from one spot – and never take more than you can eat.
- Take care where you tread so you don't damage other plants or animal habitats.
- Be careful where you pick – make sure it's not within reach of dog-walkers, or close to where herbicides might be used.
- Always wash your plants before you eat them.
- Try a small amount of the plant before you cook with it, to check your tolerance. I advise giving wild mushrooms a wide berth – especially for the novice – unless of course, you really know what you're doing.
- A good place to start foraging is on old, well-grazed pastures and edge habitats, such as woodland edges and clearings, hedgerows and pathways.
- Do a foraging course to get you started. There are many taking place around the country. Try wildfooduk.com, or totallywilduk.co.uk. For a more comprehensive guide to what's in season, check in with the woodlandtrust.org.uk.

RIVER FISHING

Do you know that fish from our rivers was once popular on British dinner tables? The Victorians couldn't get enough perch, which teemed in our inland waterways. Those not living near the sea back then relied on the abundant fresh fish from our rivers and canals. But have you ever seen perch, chub, carp, pike or the like at your local fishmonger recently? Almost certainly not.

There are many reasons for this, not least our nervousness surrounding the legality of fishing, with licences required, permissions to seek, and hefty fines doled out for those who ignore it. But it's time we rediscovered the culinary potential of freshwater fish.

First, a few rules. Go to gov.uk for information on where and when you can fish, with national and local byelaws explained, licences and permits required, catch limits, tackle and bait to use, and what you can take home. And on that note, it's a surprisingly generous allowance – from rivers each day you can take home one pike (up to 65cm), two grayling (30-38cm), 15 small fish (up to 20cm), including bream, chub, perch, tench and rudd (you measure fish from the tip of the snout to the fork of the tail). However, it's illegal to take home fish from canals, with heavy fines for those who ignore the legislation – all native fish species must be returned to the water. Non-native fish, however, must not be thrown back, so you can take them (such as zander and wels catfish) for food. Go to the canalrivertrust.org.uk for guidance.

Okay, so pike has a lot of bones – not that this stops your Eastern European angler from cooking them up for their tea, as it's a delicacy in that part of the world. Nor does it stop your Lyonnaise cook, who will tease the delicate white flakes into a featherlight pike quenelle, which is typically served with a rich, buttery Nantais sauce. But as this goes beyond most culinary skills, instead consider grayling for your dinner table.

A member of the same family as trout and salmon, grayling sports a huge dorsal fin and beautifully coloured scales, which has earned it the nickname 'the lady of the stream'. There's even a fan club, The Grayling Society, which while advocating catch and release, requests that if you do plan to eat it, "please despatch it quickly using a priest." (A priest is an angling tool for killing fish.)

Then there is perch. Not that the maximum allowed length of 20cm will yield much fish, but if you think of it like a bass-style sardine, you might want to give it a try and serve two or three per plate. Once popular in Victorian Britain, numbers had declined over the years, but now perch are back and virtually no one is eating them. Another looker, with its green scales and black stripes, it offers firm white flesh. To cook it, simply descale, fillet, toss in seasoned flour and pan-fry in butter, serving with a squeeze of fresh lemon.

Or how about chub ceviche, anyone? "It's delicious," enthused one fishing forum, though I've yet to try it. With pressure on our oceans at an all-time high, perhaps it's time to look to less familiar species from our rivers.

GROWING YOUR OWN ON BOARD

Growing food on a boat might sound a bit bonkers, but anyone who spends time on our inland waterways will have noticed the flower-bedecked boats and pots of herbs spilling out on cabin-tops, sterns and bows during the warmer months. But strawberries and salad leaves, beetroot and beans? Believe it. And really, who can resist that vibrant splash of colour and flavour that comes from tossing a handful of homegrown herbs into a dish?

Even though I have been known to create an instant herb garden on board a rented river boat for a week's holiday, this is a subject mostly for narrowboat and barge owners. The often-empty expanse of long, flat cabin roof is crying out for growbags and an array of pots. It's the perfect place for container gardening, not to mention boosting the appearance of your boat come the summer, with its handsome lush foliage – something Roka knows all about. The florist and boater, known for her business @rokabringsflowers, keeps her many followers entertained with meals on board her narrowboat *Leafy Lady* made with her abundant container-grown fruit and vegetables (see picture on page 42).

That said, what to grow and how to grow it throws up challenges that are unique to life on board, but with a little planning and a lot of patience you will reap the rewards on the plate.

THINGS TO CONSIDER WHEN GROWING YOUR OWN ON BOARD

YOUR LOCATION will determine what will thrive and what will perish. This will probably require a bit of trial and error, so expect to send a few plants to their watery grave in the early days. Start experimenting with supermarket seeds or cheap market seedlings.

USE LOW RECTANGULAR POTS, planter boxes, and trough-style tubs with a low centre of gravity and less soil than you would normally use, as it will make the pot less likely to topple over and create a mess.

MAXIMISE YOUR SPACE and increase variety by choosing plants that are happy to share a pot, such as thyme, oregano and rosemary, which are partial to well-drained soil, or mint, coriander and lemon balm, which prefer more moisture.

TRAIN YOUR PLANTS to encourage a lower but wider growth pattern, as tall plants (or pots) won't make it under low bridges or through tunnels.

SECURE YOUR POTS while on the move. Solutions range from grip mats and bungee cords, to Blu Tack and heavy-duty Velcro, depending on the size of the pot.

CREATE YOUR OWN COMPOST from food scraps, such as vegetable peelings, coffee and tea remnants, eggshells, and prunings.

TREAT PLANTS WITH NATURAL INSECTICIDES so as not to contaminate your food with chemicals (but slugs and snails won't be a problem on board).

BE MINDFUL OF THE WIND AND RAIN when considering which varieties to plant. Ensure you keep a tight watering schedule to counterbalance the drying effects of the wind.

NEVER USE RIVER OR CANAL WATER TO WATER your vegetables and herbs as it may contain waterborne parasites or contaminants. Instead collect rainwater by capturing the runoff from areas such as solar panels, or use cooled pasta or vegetable water.

BUY A PURPOSE-MADE TRAY FOR YOUR GROWBAGS to sit on, plus install a couple of strips of wood underneath each to allow for ample airflow – trapped moisture will soon lead to paint blistering and rust.

CHOOSE A SLOW-RELEASE FERTILISER that will feed your plants throughout the season and help to prevent plants from drying out.

WHAT TO GROW

Choose plant varieties that have been developed to grow better in containers and still produce generous crops while being more disease-resistant. If you are weekend boaters, like Liz and Kim, or summer boaters, then you can choose a

selection of fast-growing salad crops, herbs, and some flowers. But if you live on a boat full-time, then not a month will go by when you can't grow something.

Good for summer and weekend boaters

RADISHES can be sown from March through to September, only take 4-6 weeks to crop, and provide welcome freshness, crunch and colour to salads.

SALAD LEAVES Choose the cut and come again varieties. They will take 2-4 weeks to crop. Plant from mid to late spring through to summer to ensure a continuous harvest.

HERBS Think rocket, parsley, mint, basil, coriander, dill, rosemary, chives, thyme, marjoram, and oregano, or whatever takes your fancy.

Good for year-round boaters

RUNNER AND FRENCH BEANS Seek out dwarf or patio varieties, keeping them well-fed and watered.

TOMATOES Either start them from seed in February and March or buy as young plants in April and May, choosing tumbler varieties such as Tumbling Tom, which yields either red or yellow fruits, or try dwarf bush varieties such as Lizziana.

COURGETTES Find those developed for container-growing. You can use the flowers, too, which are great stuffed with goat cheese before frying in olive oil until golden and drizzling with honey.

Best boat flowers: geraniums, which reward with vivid colours but also ward off flies; pelargoniums will just keep on flowering; surfina petunias will keep the show going all summer; for perfumed evenings try night-scented stock and nicotiana; vibrantly coloured nasturtiums bring joy to the eyes as well as the plate when picked small and young; and almost any bedding plant will survive on a boat, though avoid begonias, which don't like the wind.

SPRING ONIONS The flavour and texture is vastly superior to supermarket versions and a little goes a long way.

GARLIC Yes, garlic is good for you, but home grown (or container-grown) garlic is even better. It's hardy and low maintenance, too.

BEETROOT Ready in as little as 40 days or once it reaches golf ball-size, beetroot handles a container well, and you can eat the leaves.

STRAWBERRIES English and Alpine varieties can thrive on a boat.

CHILLIES Particularly dwarf or compact varieties, such as Jalapeño, Habanero, Cayenne and Scotch Bonnet, though consider your desired heat level before selecting a variety…

LAVENDER It's not just a seductive scent; I add it to cakes and salads.

PEAS A freshly podded pea eaten soon after it's picked is a beautiful thing, better than any you can buy in the supermarket, easy to grow, too, with dwarf varieties available.

FLOWERS look great on a boat and add a burst of colour, but they are seasonal and will only bloom at certain times of the year, so remember to include more evergreen plants so that you have leaves all year round.

FOR THE LOVE OF HONEY

Who doesn't love honey? It's the food of the gods, literally – the Mayans worshipped Ah-Muzen-Cab, the god of bees and honey. Just as much a fan as the Maya, honey is my most scoffed sweet thing. Eaten first thing in the morning with my yoghurt, fruit and honey-sweetened granola (see page 129), added to cakes in place of processed sugar, and baked with fruit (see pages 140 and 165) as an easy dessert, or just spooned out of the jar when I need a boost of energy. I seek out locally produced jars wherever I go, savouring the array of different flavours, which depend largely on the bees' diet of nectar.

Liz makes her own wildflower honey on board. Her hive is a few steps from the mooring, hidden in a quiet leafy dell, where the bees can get on with their business undisturbed. At its most productive, when the weather plays ball, the boat hive can produce up to 150 jars per year.

Liz has been beekeeping for six years, starting first in her back garden in Peterborough, where she has two hives. Then a couple of years ago she placed an empty hive in the small woodland near the boat's mooring to see what would happen. "Within a week, a swarm had moved in," grins Liz.

How long does she spend with her bees? "I used to obsess over them, checking the bees weekly, but now it's more like 15 minutes per month. We let them get on with it and they seem to prefer it that way," Liz replies, adding that she favours a more natural approach to beekeeping.

There are two main camps with beekeeping – conventional and natural. Natural beekeepers only harvest the excess honey, and do not sell it. "It's not ours to sell," shrugs Liz. Your natural beekeeper leaves the bees alone and seldom treats them for disease, filling their hives with swarms that come in on the wing, rather than bought from dealers on the internet.

Conventional beekeepers will often treat their bees with a pesticide to repel varroa, an ectoparasitic mite that jumped from Asian honeybees to Western bees back in the 1950s, destroying millions of bees. "That felt counterintuitive to me, as the whole point of beekeeping is to keep things natural and free of pesticides," reasons Liz, who claims she finds beekeeping meditative and exciting in equal measures. "When you take the lid off the hive and they rush out into the sky, it's like a slow-motion extreme sport: thrilling and a bit scary," she grins.

Her advice to anyone considering taking up beekeeping is to get in touch with your local British Beekeeping Association (BBKA), which now boasts around 30,000 members, to attend beekeeping lessons, which many members offer, and buy yourself a hive. "Though if you ask 10 questions, you'll get 11 different answers. There's always much debate surrounding beekeeping," forewarns Liz. Her favourite books on the subject include Megan Paska's *The Rooftop Beekeeper: A Scrappy Guide to Urban Honeybees* (Chronicle Books) and *Beeswax Alchemy* by Petra Ahnert (Quarry Books).

One thing everyone does agree upon is that bees are remarkable creatures and the more you know about them, the more you want to know. Now, where's my pot of honey?

LIZ'S HOMEMADE BEESWAX BODY MOISTURISER
Melt ⅓ cup of solid beeswax in a bowl over a pan of simmering water. Add ⅔ cup olive oil and stir together until combined. Add a few drops of an aromatherapy oil of your choice and pour into a clean, wide-neck, glass container and let it cool before closing the lid.

PICKLING AND PRESERVING

Forget fancy bits of kitchen kit – preserving is the home cook's and, yes, boat cook's, secret weapon. It's the age-old way of extending the summer glut and autumn harvest into the lean winter months. And what's not to love about a glittering row of jewel-hued jars in the pantry or boat cupboard, filled with crunch and spice, and sweet and sour?

Knowing what's in season is the key – turn to the opening pages of each chapter to find out what to look for. Choose things that you particularly like and make in quantities that you can reasonably scoff (little tip: use small jars).

And before you say, "Why bother when I can buy it at the local shops?", you can't beat homemade on the taste front. Then there's the sense of satisfaction each mouthful will bring with its strong links to the past when it was all about preserving for the austere months ahead. It saves money, too, not to mention the kudos that a homemade preserve, pickle, or chutney will bring when serving friends and family. But most importantly, homemade tastes best.

JAM

There are many recipes out there for making standard jam (by that I mean equal amounts of sugar and fruit) and a few rules to follow whatever quantity of sugar and pectin you decide to use.

- Use top-quality fruit at its peak – not overripe when fruit loses its acidity, nor underripe, or bruised, all of which will affect the taste.
- It's best not to wash your berries unless they particularly need it, in which case, rinse and allow to drain on a clean tea towel.
- Peel stone fruit before using them, ditto apples and pears, which will need coring, too, of course.
- Cut large fruit into 1cm cubes, which is about right for slathering on toast.

HOW IT'S DONE

Before you dive in, here are a few simple rules and techniques that will help start you off on your pickling and preserving journey. I've focused on things that can be made easily and safely at home or on board with just the basic cooking equipment.

- Be sure to clean hands, surfaces, utensils, and produce thoroughly.
- Jars need to be hot, dry and sterile when you fill them. A dishwasher does this easily but on board a good rinse in hot soapy water works just as well, before letting them drip dry upside down on a roasting tray followed by a quick blast in the oven (about 10 minutes in the oven at 140°/fan 120°/gas 1).
- Lids must be sterilised as well, as must all the other pieces of equipment such as spoons, ladles, funnels et al.
- Place jars (minus rubber seals) and lids in a preheated oven (same as above) on a baking tray, timing it so that they are ready when your preserves or pickles are good to go.
- To seal, place the jars close to the preserving pan and using a jam funnel or a heatproof jug fill the jars with the hot mixture leaving about 1.5cm from the rim. Place the lids on the hot jars and then tighten them later once the jars are cool.
- Don't forget to label your jars!

If you are making standard jam, then almost any white sugar works. Jam sugar is the easiest to use, but the most expensive, and not necessary when you can use granulated sugar and a pectin.

Pectin is what makes the jam set. Many fruits are naturally high in pectin, from apples and plums to blackcurrants and quince. Citrus fruits are also high in pectin and acid, which is useful in the jam making process.

Fruits low in pectin include berries, such as blackberries and strawberries, and stone fruit, too, in which case you need to add pectin to enable the jam to set, which you can buy in powder form or as a liquid. Or you can combine low and high pectin fruit such as blackberry and apple to achieve a good set.

You also need acidity in jam making. You don't need to add extra acidity for tart fruit, such as gooseberries, but for most other fruit, add acidity in the form of lemon juice or food grade citric acid, which you can buy online.

And good news: you don't need much in the way of equipment for jam making. You need a pan that's about 20cm deep and wide enough to offer a large surface area (to make the boiling process quicker), a long metal spoon to stir the jam, a slotted spoon for skimming any scum that forms on the surface of the boiling jam, some clean recycled jars with lids, and a jam funnel or heatproof jug to fill the jars. That's pretty much it.

For the simplest wild blackberry jam, simmer until soft 1kg of cleaned blackberries with the juice of one lemon. Sieve to remove the pips. Reheat the pulp and stir in 800g granulated sugar until dissolved. Boil gently until the jam reaches its setting point. Test this by putting a drop on a cold plate and if it 'wrinkles' as it falls, it will set when cool. Pour into sterilised jars and seal.

For another super-quick jam, simply simmer some fruit, mix in a sweetener, a squeeze of lemon juice, and some chia seeds. This method works best with honey or maple syrup. Heat 2 cups of any juicy fruit (fresh or frozen) until it begins to soften and break down (add 1 tablespoon of water if necessary), mash with a fork then remove from the heat and stir in 2 tablespoons of chia seeds, 1 tablespoon of lemon juice and 1 tablespoon of honey or maple syrup. Leave for 5-10 minutes to thicken. Serve immediately once cooled or transfer to a sealed jar. It will last for about a week stored in the fridge.

CHUTNEYS & PICKLES

I love a pickled fig. I'm also rather partial to green tomato chutney. You can make chutneys and pickles from a vast array of fruit and vegetables. The traditional preserving agents here are vinegar and sugar, which, with salt, allow for a longer shelf life.

What's the difference? A chutney is boiled until cooked and very thick and has a relatively high sugar content. Pickles are first brined, and either placed in a jar and covered with cold vinegar, or cooked briefly and covered with hot vinegar.

Small, crisp vegetables are crucial for pickles, the brighter looking the better. I particularly love beetroot and carrot, and cauliflower and radishes.

They need brining to remove some of the water found in them to keep them fresh and maintain that crunch.

Here's a quick 'fridge pickle' recipe (they need to be kept in the fridge and last about two months). For 1kg of chopped up vegetables, make up a pickling brine with 550 ml of white wine vinegar, 200 ml of water, and a heaped tablespoon each of fine salt and sugar, plus the flavouring of your choice, such as garlic, bay, dill, or chillies.

Bring the brine to the boil; add the veg for one minute. Spoon 1 tablespoon of olive oil into each jar then divide the veg among them. Add your spices (those I like include mustard seeds, peppercorns, allspice, chilli flakes, coriander seeds, and ginger), cover with the brining liquid and seal. The pickles are ready to eat in 24 hours and will last for up to two months in the fridge. For an even quicker 'fridge pickle', see my recipe below for cucumbers.

As well as pickled figs, I love a fig chutney – especially when served with salty sheep or goat cheese. This recipe for fig chutney can be ready in under half an hour. Finely chop a red onion and put into a saucepan with 4 tablespoons of soft brown sugar, 1 teaspoon of mustard seeds, 1 cinnamon stick, 1 star anise, 1 teaspoon of grated ginger, and 4 tablespoons of red wine vinegar. Cook slowly for about 10 minutes, until the onion has softened. Add 4 figs, cut into wedges and cook for another 10 minutes, or until the figs start collapsing. Transfer to a hot, sterilised, airtight jar, seal with a lid, and store in a cool dark place or in the fridge. It will keep for 1 to 2 weeks once opened, but if properly sealed can be stored in a cool dark place for up to a year.

PICKLED CUCUMBER

Cucumber's mellow flavour makes it ideal for pickling as a delicious, crisp spring and summer treat. I love pickled cucumber with battered fish or on a burger, but it makes its most regular appearance for brunch, served with smoked salmon and a boiled egg. I love the combination of rich, smoked fish with sweet, sour, crunchy cucumber.

2 cucumbers
2 tsp flaky sea salt
1 ½ tbsp white wine vinegar
1 ½ tbsp caster sugar
Handful of dill, chopped

Finely slice the cucumbers into rounds or strips (I use a potato peeler), toss with the salt and tip into a colander. Cover with a saucer inside the colander, on top of the cucumber, and place a weight over it, such as a tin of tomatoes. Set aside for 20 minutes. Squeeze the cucumber gently to get rid of any excess water, then pat dry with kitchen paper. Toss with the vinegar, sugar, and dill. They are best eaten immediately – they can go a tad slimy if left to hang around too long.

PICKLED PLUMS

I extracted this recipe from brothers Fin and Lorcan Spiteri, who run a restaurant called Caravel and a cocktail bar called Bruno on two neighbouring barges moored on the Regent's Canal between Hoxton and Islington. They're brilliant: the floating restaurant dishes out a joyously simple menu using top seasonal ingredients, while the cocktail bar serves up imaginative twists on classic drinks. I first tried the pickled plums with their duck rillettes (see page 55), though confit duck is another popular combination, both of which I buy ready-made for on board scoffing. All you need to elevate it further is this simple-to-make accompaniment. Little tip: if you can only find small, hard supermarket plums, double the quantity of fruit and add them to the simmering liquid for 3-4 minutes to soften them up a bit.

450 ml cider vinegar
225 ml water
450g caster sugar
1 tbsp fennel seeds
1 tbsp coriander seeds
½ tbsp salt
½ tbsp black peppercorns
8 large plums

Put all the ingredients, except the plums, into a large saucepan. Simmer gently for five minutes. Meanwhile, wash the plums, halve, and de-stone. Take the pan off the heat and add the halved plums. Cover with kitchen paper and leave to cool. Remove to a sealed container, making sure the plums are covered with the liquid. Store in the fridge for 48 hours before using. They will keep for up to 6 months in the fridge.

FERMENTS

Fermented foods have been a part of the human diet for over 10,000 years. But we have embraced them in recent years, fermenting all sorts, developing an awareness and understanding of how the process affects us or, to be more specific, how it affects our gut microbiome. In short, fermented foods should be an integral part of our diets, so say the men in white coats.

The Japanese arguably eat the most fermented foods, whether to add flavour to a dish or make it central to the dish itself, with their fermented soybeans and miso-rich dishes. Korea is also at the top of that list of big, fermented food eaters, thanks to kimchi et al. But even in Europe, there's a fair amount of fermented food consumption going on, from sauerkraut and pickles, to kefir and cheese.

How can we instantly up the amount of fermented food we eat? Here are three simple ways: reach for that jar of pickles to add crunch and a kick to lunch; enjoy a spoon of kimchi alongside your cheese toastie; or why not dollop kefir on your granola for breakfast?

Sure, you can buy an array of good quality, live bacteria pickles, sauerkraut and kimchi these days, but as with jam, it's so much better when it's homemade.

For quick ferments, your jars don't need to be sterilised, but they do need to be clean; and you should use filtered water, not chlorinated, and good sea salt minus the caking agents. And trust your senses – if something smells off, start again.

For a speedy two-day carrot and cabbage ferment, mix two grated carrots and a small, shredded cabbage with one tablespoon of salt, and place in a large glass jar with a lid. Leave in the fridge for a couple of days, or longer for a more pronounced flavour.

DRIED HERBS

We dry herbs at home so why not on the boat, too? Hang them from hooks in bunches to slowly dry, scenting the air in the process. Then pack into airtight jars and store them in a dark cupboard, rubbing the herbs between your fingers straight into whatever sauce you're making. The warm, aromatic smells and flavours bring summer back in an instant.

I choose punchier herbs to dry on board, namely marjoram, oregano, thyme, tarragon, mint, and rosemary. Pick the herbs when they are at their

peak, before any flowers open. The best time to pick them is in the morning after the dew has dried. Make sure they are clean, and if not, rinse in cold water and pat dry on kitchen paper or a clean tea towel before hanging to dry for at least a week to ensure all the moisture is gone. At home I also sometimes use a dehydrator, laying the herbs in a single layer and running it for around 6 hours until the herbs are crisp and dry. Alternatively, use a cooling oven to lay herbs in a single layer on baking paper lined trays and leave overnight.

CORDIALS

Cordials date back centuries, and along with liqueurs, were first conceived as medicine. They were as likely to be found in monasteries as in apothecaries, with mixtures steeped in spices, herbs, and all manner of ingredients, to quieten nerves, settle stomachs, and generally restore good health. They never really went away, but these days you'll find artisanal cordials aplenty.

I'm not talking the kind of cordial I grew up with in the 1970s, such as blackcurrant Ribena and Rose's Lime Cordial (though I still drink the latter in cocktail classic the Gimlet – two parts gin to one part lime cordial – a good one for the boat!), but a more organic version, made without e numbers and with less sugar. A homemade cordial is an excellent alternative to booze at parties and great for making punches or classy cocktails.

Lots of fruits are suitable for making cordials – citrus in particular, especially when the fruit is at its best earlier in the year, while soft fruits give up their flavour in just a matter of days.

The most excitement though is reserved for the end of May and early June when the elderflower blooms. Is there a more evocative cordial? It's so easy to make, too, and it adds a sweetly scented aromatic note to summer drinks, in fruit salads, and in cocktails (see page 144 for my *Ellenene* Fizz).

Store cordials in a cool, dark place and make sure to always use sterilised bottles. Elderflowers are best gathered on a warm, dry day (never pick when wet) when the many tiny buds have begun to open. First check that the perfume is good, then shake gently to remove any critters.

ELDERFLOWER CORDIAL
MAKES 2 LITRES

25 elderflower heads
3 unwaxed lemons, zested and juiced
1 orange, zested and juiced
1kg caster sugar

Place the flower heads in a large bowl. Add the orange and lemon zest. Bring 1.5 litres of water to the boil and pour over the elderflowers and zest. Cover and leave overnight to infuse. Strain the liquid through a piece of muslin and pour into a saucepan. Add the sugar and the juices. Heat gently to dissolve the sugar, and simmer for a couple of minutes. Using a jam funnel or a heatproof jug, pour the hot syrup into sterilised bottles, and seal with swing-top lids, screw caps, or corks.

FRUIT LIQUEURS

Come late summer, when our hedgerows are brimming with fruit, I've got a little hooch production line on the go.

I buy cheap supermarket vodka and flavour it with wild blackberries and allotment raspberries. Ditto oranges and rum, and damsons and gin (for sloe gin see page 172). If using damsons, there's no need to de-stone them. Just tip into a freezer bag, freeze overnight before bashing the damsons a couple of times with a rolling pin. Place 500g fruit in a large jar and cover with 250g sugar and a litre of alcohol. Seal, shake and continue shaking daily until the sugar is dissolved, then stash in a dark cupboard for three months. Strain out the fruit through a piece of muslin, then bottle and hide away for at least a year, trying not to be tempted to sample it in the meantime – the longer you leave it the better it will taste.

BRITISH DRINKS

We're drinking better, we're drinking local, and we're drinking less. Things have changed enormously in recent years on the boozing front in Britain. There's now so much to choose from right here in the UK, from craft beers and artisanal spirits, to world class fizz and gut health-boosting soft drinks.

And I love that we don't have to apologise anymore for not boozing as we raise our glasses of decent-tasting, British, zero or low alcohol beer, wine or spirits. So, before you reach for that Aussie Chard in your local supermarket, consider the fast-growing range of excellent homegrown options.

BRITISH CRAFT BEER

It was all going so well for the British craft beer industry in 2019. In just over a decade, the movement (and it is a movement, so passionate are the fans) evolved from almost nothing to some 2,000 breweries around the country, becoming beacons of their community and offering a way of life that had prompted many to ditch the desk job for getting down and dirty in a mash tun.

Then Covid-19 hit. Add to that CO_2 shortages, an oversaturated market, unprecedented cost increases and slowing of sales caused by the cost-of-living crisis, and brewers then struggled to turn a profit, prompting a flurry of closures.

At the time of writing, the number of independent breweries in Britain stood at 1,815, according to figures from the Society of Independent Brewers and Associates (SIBA). You would be forgiven for thinking that this is the beginning of the end for the British craft beer industry, but you would be wrong. Because once discovered, the beauty and thrill of a craft ale is not forgotten. There is no going back.

Insiders will tell you that the UK is about 5 years behind the US in the craft beer movement. It was, after all, the US craft beer movement that largely inspired our own, and it suffered a similar blip before bouncing back. So, the British craft beer industry is sure to bounce back, too. Think about that the next time you deliberate over a pint of Easy Livin' Session Pale Ale. It's from South Wales craft brewery Tiny Rebel, by the way. And yes, it's delicious.

> **5 great craft breweries near canals and rivers**
>
> - Nene Valley Brewery, Oundle, Northamptonshire (River Nene)
> - Cloudwater Brewery, Manchester (Ashton Canal)
> - Lost and Grounded, Bristol (Kennet and Avon Canal)
> - The Five Points Brewery, East London (Regent's Canal)
> - GlassHouse Brewery and Taproom, Birmingham (Worcester & Birmingham Canal)

BRITISH CRAFT CIDER

As I'm about to share my passion for British artisanal cider, an email plops into my inbox with news of a new shop that has opened in London Fields. The Fine Cider Company is a bottle shop and tasting hub where you can discover many of the exciting new producers in the UK that have emerged over the past decade, and it's impressive.

Cider and perry have always been part of British culture, it just lost its way somewhat, swamped by big brands making overly sweet drinks and stripped of any character (or that's my biased take on it).

Happily, increasing numbers of us have discovered the joys of craft cider thanks to a growing band of small producers all over the country, from the Scottish Highlands to Cornwall, who care deeply about ancient orchards, hand-picking fruit and wild fermenting ciders. Some even use the Champagne method (which involves creating sparkling cider through a secondary fermentation in the bottle), such as Hampshire's Chalkdown Cider and Tinston in Sussex.

Add to that craft cider's impressive food-friendliness, its lower ABV, and its sustainability credentials, and it's no surprise that it's winning fans. Allow me to share some of my favourites.

Top of the list is the excellent range of ciders and perries from Little Pomona in Herefordshire. They were the first to create dry, keg-conditioned cider on a commercial basis.

This new breed of cider-maker has the freedom to experiment, which keeps things exciting. Take Hampshire's Blackmoor Estate Sour Cherry Cider, which is made by co-fermenting apples with cherries, inspired by the celebrated Belgian Kriek style. This is no longer just a movement happening among artisanal producers selling only to specialist retailers. Even Tesco is getting in on the act with its Pulpt range, made using craft principles.

This new breed of cider boasts all the complexity of wine and craft beer, complete with provenance, varietal difference, and even terroir. From the rich, earthy tannins of Herefordshire's bittersweet varieties to the clean, bright, zippy Kentish ciders, there is something to suit everyone.

ENGLISH WINE

Drive down the M20 in Kent today and you would be forgiven for thinking you were in the Champagne region of France as row upon row of vines shimmer in the breeze on south-facing, chalky slopes. With bumper crops of exceptional fruit thanks to our longer, hotter summers, production is increasing every year, helping winemakers keep up with demand from both home and abroad.

The output of the UK wine industry might still be small compared with other major wine-producing countries, but it has come a long way fast since its first commercial vineyard was planted in Hampshire in 1952. It has even won high-profile awards and trophies, mainly for its sparkling wines, which has fuelled some serious investment in land.

To put things into perspective, in 2025, according to WineGB, there were 238 wineries in the UK, with the total area under vine standing at 4,841 hectares, representing a growth of over 125% in 10 years.

"It's a success story that has shown extraordinary growth and development in the last decade as a result of significant investment," says WineGB CEO Nicola Bates.

Chardonnay leads the way in the grape stakes, followed closely by Pinot Noir, then Pinot Meunier. These grapes are often used to make sparkling wines, which in the UK accounts for 76% of all wine produced.

Add to all that continued interest from the big champagne brands, including names such as Taittinger, whose state-of-the-art visitor centre opened in spring 2025 at its Domaine Evremond vineyard near Faversham in Kent, and the future is looking bright for English wine.

5 great wineries to visit near rivers and canals

- Hundred Hills, Oxfordshire (The Thames River)
- Three Choirs Vineyard, Gloucestershire (Gloucester and Sharpness Canal)
- Ancre Hill Estates, Monmouthshire (Monmouth & Brecon canal)
- Winbirri Vineyard, Norfolk (Norfolk Broads)
- Greyfriars Vineyard, Surrey (River Wey)

BRITISH SPIRITS

The spirits market in the UK has snowballed in recent years, becoming something of a global phenomenon. British artisanal spirit brands are now proudly displayed on the gleaming shelves of top bars around the world. The UK listed 54 new distilleries in 2023, reaching a total of 387, up from 356 distilleries in 2022.

Scotch whisky still leads the way in terms of exports, and continues to thrive back at home, with whisky drinkers getting younger as they experiment with cocktails. Along with gin, whisky is our most traditional British spirit, made here for centuries, and its history continues to appeal.

Okay, so there might be a spot of gin fatigue going on as the market copes with the saturation of artisanal gin brands that have launched in recent years, but we still have a taste for it. Gin just has a bit more competition now from the likes of artisanal vodka, which is also grabbing attention, from producers such as Chase Distillery in Herefordshire, Black Cow in Dorset, and Sipsmith in London.

British rum, too, is having a moment. Although most of it is imported spirit flavoured within these shores (spiced rum sales are soaring), there are a few distilleries now producing rum from start to finish in the UK, and doing a stellar job, such as Old Salt Rum in Essex, and Sea Wolf White Rum in Scotland.

British liqueurs are making a name for themselves, too – I love London-based Kamm & Sons 'British Aperitif', and Herefordshire's Whittern Farm White Heron Cassis (excellent in a Kir Royale with an English fizz), not forgetting the oldest British liqueur, King's Ginger, first crafted in 1903 by wine merchants to the Royals, Berry, Bros & Rudd. It's the ultimate winter tipple on the water on a cold winter evening.

BRITISH CRAFT SOFT DRINKS

The low-no alcohol movement is here to stay, as we continue to see a rapid cultural shift in attitudes towards alcohol in Britain. Many of us are drinking a lot less than we used to, particularly Gen Zs and millennials, as we aspire to lead a healthier lifestyle.

Cue the wave of artisanal soft drinks hitting our shelves! You can forget the likes of Robinsons Barley Water – nowadays we want our soft drinks to deliver so much more, in terms of taste, but also with added benefits that range from improving gut health, to enhancing our mood, or even giving us better focus and energy, and with no added sugar or sweeteners in sight.

Craft-led sodas are now de rigueur in more forward-thinking eateries and retailers, with ever more interesting flavours a key demand, such as Artisan's addictively spicy ginger and lime. Meanwhile kombucha is still on the rise, from companies such as Hip Pop with its popular kombucha soda, with lots of different flavours on offer. British-made milk-based drinks are also on a roll, with kefir and milkshakes dominating our shelves, especially those boasting immune support.

Forget the overly sweet supermarket own-label mixers. Since Fever Tree paved the way a couple of decades ago, it's now all about artisanal mixers – think blood orange and elderflower tonic from the London Essence Co, or cucumber and watermelon tonic from Double Dutch. Posh pop, too, is in the spotlight, with seasonal flavours in vogue from the likes of Luscombe, Cawston Press and Franklin & Sons.

Soda water is no longer handed to you in a standard glass still warm from the dishwasher at your local pub. Now it's all about 'craft softs', as the industry calls them, from Soda Folks and Redemption, and Fhirst's Living Soda, packed with plant fibre and active cultures. Or how about Apeal World's Activate ACV range, which has a sparkling apple cider vinegar base? Not to mention the multitude of soft drinks incorporating CBD and vitamins. It's a bandwagon our bodies might approve of, at least.

RIVER RESCUE

Rant alert! The water quality of our rivers is a travesty. Not a day goes by without news of the latest river pollution drama, poisoning both wildlife and wild swimmers, and upsetting delicate ecosystems. And according to the experts, it's only going to get worse as our climate changes.

The Rivers Trust – an umbrella organisation for 65 individual river trusts across Britain, Northern Ireland and Ireland as well as being river catchment conservation experts – found in its latest report (2024) that none of England's rivers are in good health.

Sewage is largely to blame, both treated and not. The amount spilling into England's rivers and seas doubled in 2023, with 3.6 million hours of spills compared to 1.75 million hours the year before, according to the UK's Environment Agency. Our antiquated sewage systems simply can't cope in heavy rain, is the collective shrug of water company CEOs and shareholders. Add to that a growing population and expanding infrastructure, and it's a perfect s**t storm. Literally.

But it's not all doom and gloom. There are various exciting research projects underway that are looking at ways to improve water quality. Projects such as PACIFIC (Pathways of chemicals into freshwaters and their ecological impact) focus on freshwater microbial communities. These play a crucial role in maintaining the health of freshwater ecosystems and the organisms that live in them. Research is being led to find new ways of lessening the impact of micropollutants, aka household chemicals, in this ecosystem.

Then there are the more holistic projects, such as the one that introduces new wildlife-rich wetlands near sewage works as a natural filter, encouraged by conservation charity The Norfolk Rivers Trust. These 'integrated, constructed wetlands' (ICWs) effectively fine-tune the water before it gets into the rivers. They are working so well, in fact, that a growing number of water agencies are now building these wetlands near their sewage works.

And we can be thankful for other initiatives in place, such as Riverfly, which keeps an eye on our river health by looking at the mix of invertebrate species in the water. These indicate the impact of pollution, and the data is then fed into a national database.

There is no doubt that these combined projects are making a significant impact on national river health, but we can play our part, too (see page 70 on sustainable boating).

5 great books to read about our inland waterways

- *Narrow Boat* by L.T.C Rolt (The History Press)
- *The Canal Guide: Britain's 55 Best Canals* by Stuart Fisher (Adlard Coles)
- *Water Gypsies* by Julian Dutton (The History Press)
- *Tales from the Tillerman: A Life-long Love Affair with Britain's Waterways* by Steve Haywood (Adlard Coles)
- *Three Men in a Boat* by Jerome K. Jerome (Wordsworth Editions)

5 great river and canal trips for beginners

- **LLANGOLLEN CANAL, WALES** A border-crossing canal that links England and Wales, with stunning countryside and a few engineering feats.
- **ASHBY CANAL, LEICESTERSHIRE** Apart from not having to deal with any locks for its entire 22-mile length, this is a gentle, pretty route for first-time boaters.
- **NORFOLK BROADS** Choose the 26-mile run from Brundall to Beccles with only a couple of swing bridges and a chain ferry to negotiate, and birdlife galore.
- **WORCESTER AND BIRMINGHAM CANAL** No locks between Alvechurch and the W&B Canal, and city centre moorings in Birmingham's iconic Gas Street Basin.
- **MONMOUTHSHIRE AND BRECON CANAL** There are very few locks along its 35-mile length plus mountain views and starry skies that will lift the spirits.

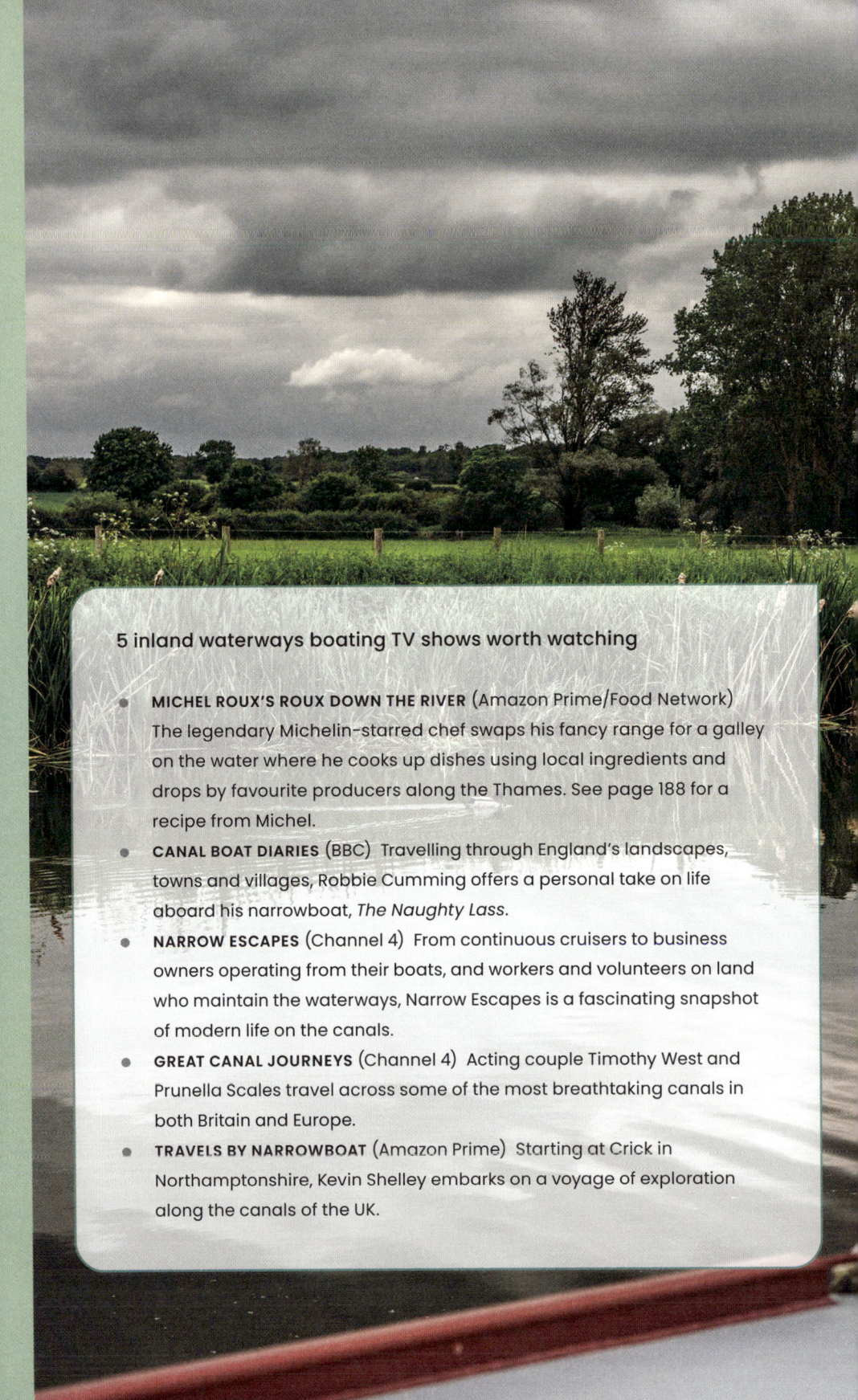

5 inland waterways boating TV shows worth watching

- **MICHEL ROUX'S ROUX DOWN THE RIVER** (Amazon Prime/Food Network) The legendary Michelin-starred chef swaps his fancy range for a galley on the water where he cooks up dishes using local ingredients and drops by favourite producers along the Thames. See page 188 for a recipe from Michel.
- **CANAL BOAT DIARIES** (BBC) Travelling through England's landscapes, towns and villages, Robbie Cumming offers a personal take on life aboard his narrowboat, *The Naughty Lass*.
- **NARROW ESCAPES** (Channel 4) From continuous cruisers to business owners operating from their boats, and workers and volunteers on land who maintain the waterways, Narrow Escapes is a fascinating snapshot of modern life on the canals.
- **GREAT CANAL JOURNEYS** (Channel 4) Acting couple Timothy West and Prunella Scales travel across some of the most breathtaking canals in both Britain and Europe.
- **TRAVELS BY NARROWBOAT** (Amazon Prime) Starting at Crick in Northamptonshire, Kevin Shelley embarks on a voyage of exploration along the canals of the UK.

SUSTAINABLE BOATING

There's no doubting the sense of connectivity we all feel living closer to nature when we spend time on the waterways. The wildlife alone will excite – from kingfishers and otters, to grebes and herons. You can also take comfort in the fact that your carbon footprint is greatly reduced, as boaters tend to use less water and energy. But there is always more you can do to make your boating life more sustainable, whether you are a weekend cruiser or a full-time liveaboard.

8 WAYS TO MAKE YOUR BOATING LIFESTYLE MORE SUSTAINABLE

1. RECYCLE, UPCYCLE

Taking control of your rubbish can make a massive difference to your environmental impact. Before you throw something in the bin, ask yourself, could it be repurposed? For example:

- That yoghurt pot could be put to use as a container for screws and nails.
- Save jam jars for shaking up salad dressing.
- Use empty loo rolls as extension cord organisers.
- Cut a plastic milk container in half and use as a scoop for pet food or potting soil.
- Use kitchen paper roll as a (reusable) plastic bag store.
- Loo rolls can be used for seedlings and fire-lighting.
- Use knitted washing up cloths so they don't leave fibres in the water.
- Cut in half an old orange juice carton then wrap it in fabric and use as a planter for herbs and flowers.
- Use old (clean) socks to dust off your window blinds.

For more recycling inspiration, search online. Failing all that, if your mooring doesn't offer recycling facilities, you can use online tools such as recyclenow.com to help you find the nearest recycling bank – simply type in the town or village where you are moored and select the nearest places to recycle.

2. CLEAN GREEN

It's not just rubbish that can cause problems for our waterways. Cleaning products, for both body and boat, can be incredibly toxic and harmful to the environment. For most boaters on the canal, grey water drains directly into the waterways, potentially harming wildlife and upsetting delicate ecosystems, so it's vital to use products for washing, washing up, and laundry that are not only eco-friendly, but are safe for the waterways. Regular soap, for example, contains compounds such as sodium lauryl sulphate (SLS) that can be toxic for wildlife, while phosphates in standard cleaning products encourage the growth of duckweed and pennywort during the summer months, which can cause problems for both boaters and wildlife, creating thick mats on the surface of the water, clogging boat engines. I make my own cleaning spray – combining white vinegar, baking soda, lemon juice and essential oils, such as cleansing, fragrant eucalyptus and lemongrass.

3. A LOAD OF RUBBISH
I'm sure you would never lob anything into the water, but sadly a small minority do, so keep a net on board to scoop out bits of rubbish you might see when out motoring. The Canal & River Trust (canalrivertrust.org.uk) encourages anyone who wants to help to join its Plastics Challenge. "If everyone who visits one of our canals or rivers picks up just one piece of plastic and takes it home, they'd be clean within the year," it dares.

4. ENGINE MAINTENANCE
A perfectly working engine is more efficient and produces fewer emissions – fact. So learn how to keep your engine in prime condition. It might sound daunting, but most engine services are easy to DIY and there are plenty of courses out there to help you.

5. KEEP YOUR DISTANCE
From wildlife, that is. Yes, it's the raison d'être for many boaters – is there anything more mindful than soaking up the sounds of nature at play? But it's important to remember that this is their world, too. Getting too close can disturb their natural routines. Wildlife deserves space and respect.

6. A QUICK WORD ABOUT POO
For the ultimate sustainable loo, look to the compost toilet. They are an increasingly popular addition for your eco-conscious boater. But even just changing to greener toilet chemicals for pump-out and cassette toilets will help you do your bit for the environment.

7. SUN POWER

It's the norm these days to see solar panels on boats. There are very few boats now that can't create their own electricity, either by harnessing the sun's energy or relying on the wind (or both, more on that below). In fact, there are now narrowboats being built that are entirely electric, their energy drawn from a bank of solar panels covering the roof. Southwest London-based Thames Solar Electric is the UK's first supplier of solar electric boats, which boast large battery banks so you can motor along (silently), even in the winter. Their super-smart wide beam canal boats have now hit the waterways from the Droitwich Canal to the Great Ouse. And remember, electric boats get a 25% discount on their licence fee – just saying. Though having solar panels isn't enough – you need to keep them clean and out of the shade to ensure optimum performance.

8. WINDY BUSINESS

Second to solar panels are wind turbines. They're easy to mount and can help top up batteries on a cloudy day. The downside is they are a tad pricey and rather noisy, though still way better than hours of running noisy, polluting diesel engines and generators. You could also look at ways of reducing the number of things that need powering by electricity – do you really need a hairdryer, or a microwave? Swap halogen lighting to LED, add a few solar lights, try a wind-up radio, and even consider reverting to a basic phone (hello, old friend, Nokia 3210).

BOAT RENTAL

Boating on our inland waterways has never been so popular. We live, holiday and weekend on boats and we can't get enough boating on telly (watching Robbie Cumming pottering along canals in his narrowboat *Naughty Lass* in his *Canal Boat Diaries* for the BBC is an instant stress buster). In short, we're rediscovering the country via our inland waterways and finding a much-sought-after sense of peace and liberation along the way.

You might be thinking about living on a boat but want to rent one for a stint to try out the lifestyle first, though renting, frustratingly, is not as easy as you might think, as there are few residential moorings available (visit waterways.org.uk for more information on the subject). If that's your mission though, you will need to do your sums – mooring fees can be expensive.

Or you might own a boat to retreat to during the weekends, or are planning to holiday on a boat. In fact, many of the former do the latter as a way of further exploring the British countryside.

Aside from experiencing the country in a completely different way it's the sense of calm that ensues on a boating holiday that is a major appeal, contributed to in part by the unique biodiversity that thrives along our canals and rivers thanks to the connectivity of the network. Plus, the freedom to travel from town to village to city with its offer of different things to see and do every time you moor up in a new spot becomes addictive, as any seasoned boater will tell you.

Today there are more than a hundred hire firms operating on Britain's inland waterways. So pick a spot you like the sound of, then look for hire boats in that area, as we did recently on the Norfolk Broads. And you can forget emptying loos on your holiday hire boat – today's holiday rental vessels have all the comforts of home (and more in some cases, see below), from showers and flushing loos to central heating and full-size cookers, microwaves and TVs. Add to all that the knowledge that you are contributing to the maintenance of the waterways network as well as supporting local waterside businesses, and things are looking rosy for British rivers and canals.

A WEEKEND ON THE NORFOLK BROADS

A cormorant perches on a branch, wings outstretched, drying off after a hard day's diving on the River Yare. The water is still – only the wake of our cruiser disturbs the glass-like surface as we motor upriver while the sun slips behind

the trees. Moments earlier, a flash of blue caught our attention as a kingfisher streaked low over the water before disappearing into the reeds. Then a great-crested grebe sidled up imperiously, only to dismiss us with a ruffle of its *Peaky Blinders*-style plume.

Just half an hour in and the Norfolk Broads is already working its magic. The anticipation builds for the weekend ahead, starting with a pie and a pint in a pub with its own riverside mooring. This will kick off the first of three nights in our lavish floating home, allowing us to experience a Norfolk we have never seen.

All mod cons – and some

Meet *Evita*. She is the flagship of five luxury river cruisers owned by NYA Private Charter, based at the Brundall Bay Marina near Norwich. There's much cooing over the two master en-suite cabins, both beautifully appointed, with high thread count sheets, sumptuous smellies, and even TVs. Plus, there's a galley that's smarter than our kitchen at home, a white leather sofa-wrapped saloon with surround sound, and a vast flybridge complete with drinks fridge for ultimate in-boat posing.

We're lucky, the forecast is looking good for our long weekend on the Broads on our large luxury cruiser, the size of which initially alarmed, though the blurb assured that *Evita* is manageable even for a novice boater. While I'm more used to yachts, two of us are seasoned narrowboaters on non-tidal rivers (yes, it's Kim and Liz) but we reckon that between us we should be able to work it out. Still, we pay close attention to the charter company's instructor, Gary, who takes us through the safety briefing and demonstrates the bow and stern thrusters, which will make mooring a breeze, he promises.

A pint and a pie

It's not long before we get to use those thrusters, as they slide us neatly onto our first mooring, at the Water's Edge Restaurant & Bar in Bramerton. Its raffish proprietor, who everyone calls Cookie, grabs a rope to secure us, leaving enough slack to compensate for the drop in tide. There are two daily tides on the Broads, rising and falling twice in every 25 hours or so.

As dusk settles and the temperature drops, we prop up the bar with a pint of Woodforde's Brewery's excellent amber ale Wherry, choosing the pie of the day for supper, which today is pork, cream and mustard. Then it's back to our glamorously lit boat and its swish saloon for a raid of the well-chosen, locally sourced welcome hamper.

As we sip glasses of Beeble Honey Whisky and nibble on cherry and almond cake from the Norfolk Cake Company, our Spotify apps synced with the boat's Wi-Fi, we're feeling thoroughly spoiled.

It's not the sound of birds chirping and reeds rustling that rouses the cabins' occupants the next morning, but a rowing boat's cox barking "Heads up, keep long!" at his crew of four as the boat zips through the dawn mist, sending gentle ripples our way.

Destination Suffolk

Today's plan is to get to Beccles. I have an ulterior motive in suggesting this destination, which will become clear later. But for now we have a route to plot, which includes dodging a chain ferry and a multitude of tacking dinghy sailors (power gives way to sail), plus correctly navigating a fork in the river. And there is always the tide to watch, especially at Reedham, where a quick call to the station master might be needed to open the swing bridge so we can pass through with our flybridge canopy intact. Though we needn't have worried as the boat is a joy to handle, neatly skirting dinghies and paddleboards, and holding steady while waiting for the chain ferry to dock.

If we'd had more time, we would have moored at Reedham and walked a section of the Wherryman's Way there and back to Polkey's Mill, built in 1880 to drain water from the neighbouring marshes on the River Yare, and one of 21 buildings looked after by the Norfolk Windmills Trust. And we would have made time to tour the 10-barrel Humpty Dumpty Brewery, stocking up in the shop after (open every day from 12-5pm). But with four more hours of motoring ahead we give the picturesque river frontage a wave and press on.

A patrolling Broads Ranger nods a greeting at the entrance to the River Waveney, our cue to fork right. The river narrows here, and the banks are teeming with life. Herons are perched on fallen branches scanning the water for their next mouthful, and birds of prey on a fly-by are doing the same. The maximum speed drops to 4mph so it's a good moment to grab lunch, which we eat up on the poop deck as we've christened the flybridge – sharing a couple of bottles of Wherry and a selection of Norfolk cheeses picked up the day before from Blofield Farm Shop near the marina.

Included is Becky Enefer's excellent ewe's milk brie-style cheese, Norfolk White Lady, and one of my all-time favourite British cheeses, Baron Bigod from the Fen Farm Dairy as well as a couple of cheeses from the redoubtable Mrs Temple, Binham Blue and Walsingham Smoked, with a spoonful each of Candi's Bramley Apple and Norfolk Ale chutney to add contrast.

A quick call to Beccles Yacht Station establishes the location of our pre-booked mooring (a must in the height of the season), and we soon clock our spot at the edge of the basin where harbour master Martin is waiting and ready to receive our ropes.

In Beccles, we climbed to the top of the belltower of the 1500s-built St Michael's church to get the lay of the land where the river acts as a boundary between Suffolk and Norfolk. One of the gateways to the broads, Beccles is a busy market town dating back to the Anglo-Saxons.

Armed with shopping baskets, we leave the boys to their football and pints in The Bear & Bells, and my mission is unveiled – a trip to artisan bean-to-bar chocolate maker Harris & James, which I first fell for in Aldeburgh (where it has another shop) a couple of years back.

Weighed down by a couple of boxes of its irresistibly thick dark chocolate-covered shortbread and tanked up on its house blend along with samplings of its innovative bars, we make tracks to nearby greengrocer The Pavilion for dill, chives, fennel and citrus, and pick up fat trout fillets from The Fishermonger at No.4 for supper back on board (for the recipe, see page 136). To drink with it? Norfolk's Winbirri Vineyard Bacchus, which famously scooped Best Single Varietal in the World at the 2017 Decanter awards. The vineyard is a short drive from Brundall Bay Marina, so it made a logical stop en route.

A chilly morning greeted us on our last full day, so there's no rushing from our luxe, toasty lodgings. Instead an illuminating hour is spent watching magnet fishing from the bank. "They've pulled up all sorts, from bicycles to a First World War gun. People complain, but it's environmentally friendly, isn't it?" reasons harbour master Martin, as we settle our 'side on' mooring fee (£20).

There's plenty more fishing to see on our return journey back to Brundall, as solitary anglers perch on specially adapted chairs, which appear among the reeds at regular intervals along the River Waveney.

Fishy business

People have always fished the inland waters of the Broads and, up until the 19th century, it was a big business. Smelts were particularly popular, usually fried in breadcrumbs, but perch was also sought-after, while pike was (and still is) considered the prize catch. Tastes have changed and these days coarse freshwater fish is no longer widely eaten in this country. Plus, the authorities don't recommend eating fish caught in the Broads, but should you wish to do so, just follow the local rules and guidance on size and catch limits (see page 38) and be aware that the Broads coarse fishing season runs from 16 June to 14 March every year.

We reach Coldham Hall Tavern and its riverside mooring just in time for last orders and its stellar Sunday roast, the pub conveniently positioned opposite the marina where we are due early the next morning when we will have to (very reluctantly) return *Evita*. It's been beautiful, Norfolk Broads. Delicious, too, as the region brims with edible innovation, making it a boating holiday to remember.

Want to book *Evita*? Then visit hoseasons.co.uk using reference BH2740. Expect to pay from £1,785 per week, though three-night breaks are also available.

NARROWBOAT HERITAGE

Meet *Ellenene*. It was love at first sight for Kim and Liz. Add to that the narrowboat's permanent mooring on a tranquil spot on the River Nene, just a short cycle from Oundle with its well-stocked shops, and they had found the perfect bolthole to escape city life.

Built in 1995 and measuring 55ft x 6ft 10in, with its steel hull and superstructure powered by a diesel engine, *Ellenene* is named after Kim's mother. So yes, the couple did defy deep-rooted superstitions and rename their boat (which some believe might bring misfortune to a vessel), but it's not so different from the original *Emmeline*, they reasoned.

They've since given *Ellenene*'s exterior a makeover, courtesy of wonderfully named narrowboat painter Ptolemy Lane, choosing a smart new burgundy and grey livery. And having explored the length of the River Nene, they have big plans to travel the wider canal network now that Kim has retired from his job in telecommunications. "You can get onto the canal system at Northampton from The Nene and then we can go north or south," he enthuses.

While many of today's narrowboats boast mod cons *Ellenene* does not. Hers is a compost loo, a cosy wood-burning stove, a serviceable shower, and a cool box in a compact galley that works well for their weekends on board. "And that suits us just fine," the couple grin.

NARROWBOAT OR NARROW BOAT?

Purists will tell you that a narrow boat refers to an original or a replica, and a narrowboat (no space) is a modern boat used for leisure or as a home. Just never call a narrowboat a barge, which is a much wider, cargo-carrying boat or a modern boat modelled on one.

INLAND WATERWAYS RENAISSANCE

Britain's canals and rivers have never been so busy. *Ellenene* is one of some 35,000 boats (according to the latest boater census conducted in 2022 by

the Canal & River Trust) currently cruising Britain's inland waterways, from renovated working boats to swanky new cruisers, hire boats to hotel boats. Once a catalyst for industrial revolution, now canal boats are a catalyst for inner city regeneration.

The Romans used barges to carry grain and other supplies pulled by teams of men or rowed by standing oarsmen, digging channels to connect one river to another. In the Middle Ages, some rivers were improved, and short canals were dug to carry stone for building churches and cathedrals.

GET KNOTTED

Usually made of cotton for its good stretch, rope was used for a variety of purposes on canal boats, from towing and tying up to securing cloths over cargo and for decoration, such as a 'Turk's Head' on top of the rudder, with a variable number of interwoven strands forming a closed loop.

Then long, thin wooden boats were built to go directly into mines on underground canal systems to carry cargo. Thanks to canal engineer James Brindley, these evolved into the first narrowboats pulled by horses, the way to transport goods from the late 1700s right through to the Second World War. In fact, Brindley played an essential role in shaping the way our canals were built.

Did you know that narrowboats are unique to Britain? The shape and size of the narrowboat as we know it today – less than 7ft wide and no longer than 70ft – is down to Brindley, who reached an agreement with the Trent & Mersey Canal Company in 1766 to build the first lock at precisely 7ft x 72ft (to save water and costs), along with subsequent tunnels and locks when they expanded the network beyond the Midlands.

Brindley's achievements were impressive, from creating the longest man-made tunnel in the world, the 8,778ft Harecastle Tunnel, which takes 30 minutes to motor through, to the ambitious Barton Aqueduct on the Bridgewater Canal, the world's first navigable aqueduct. He even has a square named after him, Brindley Place, in Birmingham, which lies at the heart of the country's canal network.

As the railways developed and economic pressures started to bite, boatmen's families moved onboard, often towing another boat behind to create more space for living and cargo, known as a 'butty'. In the 1870s, more than 40,000 people lived on the boats they operated.

ROSES AND CASTLES

The origin of the wonderfully elaborate decoration on traditional canal boats is a bit of a mystery. No one seems to know exactly where it came from. Some say it's Gypsy in origin, even though there isn't a strong link to this community (whatever the popular TV show *Peaky Blinders* might suggest). Others say it's reminiscent of Scandinavian folk art, or 18th century Dutch sailing barge design, which were decorated in a similar way, with roses and castles a recurring theme.

The why is easier: narrowboats were famous for their space-saving ingenuity and for their cosy interiors – especially once families had come to live onboard showing off their quarters with gleaming brass, delicate lace, and decoratively painted housewares.

The most elaborately decorated boats back in Victorian times extended their designs to the exteriors, adding ornate lettering with the boat's name and owner, and even decorated the equipment itself, from the rudder to the gang plank supports. The practice declined as canal life dwindled, but it has seen a revival in recent times – thanks to the increasing number of pleasure boats and the huge pride people take in them – albeit these days it's more likely a computer-printed transfer than a hand-painted design.

But by the end of the Second World War, as road transport now ruled, many narrowboats had been abandoned. Canals suffered from a lack of maintenance and even closure, and cargo-carrying became almost non-existent. The future was looking bleak.

Then along came the publication of *Narrow Boat*, a wildly popular book about life on English canals (still in print). A meeting between its author, L. T. C. Rolt and one Robert Aickman, led to the formation of The Inland Waterways Association in 1946, convincing the authorities that our canals needed saving. The government listened and in 1948 the waterways were nationalised – cue the pleasure boater.

By 1950, 1,500 licences had been issued for canal cruising. By 1960, that figure had increased to 10,500 licences. Then in 1968, Transport Minister Barbara Castle orchestrated a new act that sought to retain and restore British canals.

The 1970s saw a growing number of companies offering boats for holiday rental, with new hire fleets being built and boatyards busy once more, starting first in the Thames and the Norfolk Broads, and then expanding to other waterways.

Fast forward to the last decade, which saw cash-strapped creatives unable to get on the property ladder discover a new kind of des res – the houseboat. There are now some 3,000 houseboats in London alone, twice the number of eight years ago.

The pandemic then prompted another wave of love for our waterways (for me, certainly), at a time when overseas travel was curtailed or restricted, when river and canal life became a restorative tonic for our troubled times.

Since 1946, major links in the canal network have been revived, including the Kennet and Avon Canal between Reading and Bath, and many other schemes are now nearing completion. The future is looking rosy for British inland waterways, and for narrowboat life in general.

A FOODIE WALK IN WINTER ALONG THE REGENT'S CANAL

Any canal enthusiast worth their salt will walk The Regent's Canal. Winding from the Paddington arm of the Grand Union Canal at Little Venice to Limehouse Basin, it's narrowboat central in the UK, with many of the 3,000 or so boats registered as primary homes in the capital right here on this 8.6 miles stretch of waterway. And you can see why, it's right in the heart of arguably the world's most thrilling capital city, with its exciting, diverse dining scene. So, with empty bellies we set out one chilly December morning to explore a part of London that's less familiar, guided by our appetites, starting in Mentmore Terrace in Hackney's London Fields.

This is the home of the E5 Bakehouse, a bakery, café, flour mill and shop, which first opened in 2011, driven by a sustainable approach and a love of ancient wheat. Folks gather to munch on its stellar baked goods, so we start the day with a feta and za'atar Danish, and a chocolate and walnut babka. We also leave with a loaf of its signature Hackney Wild sourdough and a selection of British cheeses to try back at home, adding a bottle of North Herefordshire Little Pomona Brut de Poire to wash it all down from The Fine Cider Company next door.

It's just a 10 minute walk to the Regent's Canal via Broadway Market, with its busy vendors, stylish bookstores, artsy cafés and posh food shops, such as top butcher Hill & Szrok with its excellent fennel and garlic sausages. And it's impossible to walk past 58 Saray Broadway Café without dropping in for a Turkish gözleme, a 2ft wide crêpe cooked in the traditional way, ours stuffed

with spinach and cheese. Hang a right on the canal and eventually you'll hit Granary Square near King's Cross, our destination. But first, we're keen to check out the 878m Islington Canal Tunnel, which narrowboats of old took an hour or more to pass through by 'legging' – boaters used to lie on wooden planks and 'walk' their vessels through using the tunnel wall.

These days it's about a 15 minute chug with a diesel engine but with no towpath on this stretch, we climb the steps to weave through Islington before picking up the path again near Regent's Wharf.

Before all that though, it's a quick visit to the Spiteri brothers' acclaimed barge restaurant, Caravel, just before Packington Bridge. We've left room for a starter and choose duck rillettes with pickled plums, which were so good that I prised the recipe from chef Lorcan Spiteri (pictured right, see page 53), thinking that they would be perfect served onboard *Ellenene* with a tin of confit de canard, the plums cutting through the richness a treat (it is).

We continue along a stretch of the canal where boats are now routinely moored two abreast, reaching The London Canal Museum where we've pre-booked a Mince Pie Cruise to get us in the Christmassy mood.

The museum is worth the trip alone, of course, offering a fascinating insight into the heritage of Britain's waterways. But adding a cruise to the package – this one through the Islington tunnel on the museum's fairy light-lit, tinsel-decked narrowboat – steals the show, as we munch on excellent Sunflour Bakery mince pies while listening to our guide, Andy, paint a vivid picture of Victorian canal life.

We complete our canal walk at Granary Square with a browse around the world's best barge bookshop Word on the Water, finishing with a warming mulled wine at Vermuteria in swanky Coal Drops Yard, all the richer for canal history, appetites well and truly sated. If we'd had time, we would have continued along the canal for another hour to Paddington and The Cheese Barge, to share a whole Baked Baron Bigod cheese, scooped up with Coombeshead sourdough and garlic roast potatoes. Another time.

BOAT BARBECUES

The Brits aren't generally great at barbecues. All those years haunted by burnt sausages and undercooked chicken – or maybe that's just my friends and family? But these days, helped by some five-star bits of kit, such as ceramic charcoal grill brand Kamado Joe (I've got one at home, they're fab), and taking inspiration from the growing number of chefs who have made cooking on fire their thing, we're all getting much better at it.

Okay, so we're still a long way off the barbecue fanatics that reside in the southern states of the US, where barbecue is a revered cultural experience, with barbecue restaurants in every town and barbecues in every backyard. Real barbecue isn't grilling stuff, an American pit master will tell you. They're talking low and slow, which means cooking over an indirect heat, similar to smoking. Cooking over a direct heat is actually grilling – fast and hot – they'll sniff. Though, they concede, grilling is just fine for burgers, steaks and chicken breasts. But as grilling is all most of us in this country will ever do, on or off a boat, let's embrace the fast and hot.

Which barbecue to choose? Well, it has to be small, so when not being used it can be placed in a bag and stored in a locker under seating, for example. Consider a suitcase style barbecue, which folds down easily for storage. Don't even think about using a disposable barbecue – they're full of nasty chemicals, make the food taste bad, and they can't be recycled. When using while boating, make sure it's far enough from the boat so any hot embers blowing in the wind can't reach the boat furnishings or anything else that can burn. I'm sure I don't need to say this but just in case – it's never safe to have a lit or cooling barbecue in a cabin or covered cockpit area.

A few barbecue pointers:

- Before you start, get that grill clean, brushing away any charred bits.
- Keep food from sticking by rubbing the grill with vegetable oil.
- Wait until the coals are a uniform ash grey before grilling.
- Leave space around each food item to allow for even grilling.
- Avoid pressing burgers with a spatula as they grill because this causes them to lose their tasty juices.
- Turn meat just once on the grill.
- Keep a spray bottle of water handy in case of flare-ups.
- Store raw meat and fish separately before cooking, and use different utensils, plates and chopping boards for raw and cooked food.

- Never wash the meat or fish – it just splashes germs.
- If you're worried about whether the meat or fish is cooked properly, use a thermometer.
- Always let meat rest for a few minutes before you eat so it's nice and tender.

MEAT BARBECUE

So, what to grill beyond burgers and sausages? How about honey-mustard-marinated pork tenderloin, grilled venison with a fennel seed and peppercorn rub, or jerk-marinated chicken? A marinade will not only give your meat a big hit of flavour but it will also tenderise it. There are hundreds of marinade recipes out there but in my view it's best to keep it simple – why mask the flavour of a perfectly good piece of meat?

For chops and steaks, try combining rosemary, thyme, red wine vinegar, honey, olive oil, and crushed garlic. See page 138 for my all-purpose meat marinade for barbecues.

Or try this super simple, but very effective marinade for skin-on chicken breasts: make three or four shallow cuts in each breast, mix together the juice from a couple of lemons with a tablespoon of freshly ground black pepper and rub into the chicken. Place on a dish, cover with cling film and leave to marinate in the fridge for four or five hours, then grill skin side down for six minutes or so before turning and grilling for a further eight minutes, or until the juices run clear. Rest for 10 minutes then dust with a little celery salt.

HOW NOT TO BURN YOUR SAUSAGES ON THE BARBECUE

Consider boiling your sausages first. If it's good enough for the Aussies, who often buy them precooked this way for their barbecues, then it's good enough for us. Little tip: it works better for fat butcher sausages. Bring a large pan of water to the boil. Separate the sausages, plunge into boiling water, and cook for about five minutes to firm them up. Drain off the water and leave to cool in the strainer. For grilling, rub the precooked sausages in a little olive oil and grill over medium coals turning frequently until golden brown.

FISH BARBECUE

I love barbecued fish. It imparts a seductively smoky flavour and delivers satisfyingly crispy skin. But it's easy to overcook fish and rip said skin, so yes, barbecuing fish can be a little intimidating. But armed with some knowledge and a fresh fish, anyone can do it.

First, choose a fish that has a good amount of natural fat. The oilier the fish, the less it sticks – think sardines, trout and salmon, tuna and swordfish. For beginners, choose meaty, firm fish steaks at least 1.5cm thick.

Then oil your fish, using olive oil, which encourages caramelisation and prevents sticking, and lightly season. Next, grilling. The key here is patience, that means not checking every minute if your fish is stuck - let it crisp up and seal first, then use a fish slice to roll it over gently. For peace of mind, use an oven thermometer – cooked fish should have an internal temperature of at least 63°. Once cooked, I love to serve it with a punchy sauce, such as a sauce vierge.

Prawns on the barbie couldn't be easier, plus they look great and taste fabulous. Buy deveined large uncooked tiger prawns, then marinade for an hour in garlic, lemon, olive oil and parsley before grilling over medium coals for five minutes a side, basting as you go. Then slather with garlic and parsley butter to serve.

Sardines are the perfect fish to barbecue, small enough to cook through and oily enough to baste themselves. Just use a grilling frame. Rub them in oil and arrange in the frame, after 4 to 8 minutes on the grill, flipping halfway, they're ready to eat with a little seasoning and a squeeze of lemon.

Or go big with a whole sea bass, mackerel, trout or bream, which can be cooked directly on the grill bars (or for ease, in a fish basket) over glowing coals with thick slices of lemon. Remember, wait while the skin crisps and don't fiddle – about 10 minutes for a 2kg fish on one side, with a further eight minutes on the other side (or five minutes a side for mackerel), then season and serve.

HOW TO CLEAN A BARBECUE

Regular cleaning of your barbecue is not only important for prolonging its life but it also helps to prevent the spread of food poisoning. When it's sufficiently cool, remove the grill and soak in soapy water before scrubbing hard with a non-metal brush.

VEGETABLE BARBECUE

Vegetables can be just as delicious and filling as any meat and significantly cheaper if you stick to what's in season. A favourite is sweet-fleshed red peppers, halved, seeds scooped out, placed on an oiled sheet of foil. Put a couple of anchovy fillets and a splash of oil from the tin in each pepper, add a few slivers of garlic and a splash more olive oil, draw up the foil and twist it shut, then repeat for each half pepper. Cook for half an hour or so on the edge of the grill where the heat is gentlest until soft.

I also love barbecued aubergine, soft and unctuous, drizzled with tahini. Or try dried mint-marinated halloumi on a skewer with courgettes, and dress barbecued fennel with a black olive tapenade. I'm a fan of barbecued chicory too – the bitter leaves sing with the charring from the grill – cut in half lengthways, rub with olive oil and grill cut-side down until you get charred grill marks, flipping over for a couple of minutes on the other side before removing, seasoning, and drizzling with a little more olive oil to serve.

And I can't wait until whole, fresh sweetcorn starts appearing on greengrocers' shelves. Prepare by pulling out the silky threads, keeping the green leaves intact, and twisting the leaves shut at the pointy end with a small length of garden wire. Bury them in the embers of a fire for half an hour or so before digging them up, discarding the charred leaves and slathering in butter, salt and pepper.

Asparagus also loves a char-grill. The taste intensifies, while the texture remains crunchy. Trim away the woody stems and oil the spears. Cook on the barbecue for 5 to 10 minutes or so, turning as necessary, then serve with parmesan shavings and a grind of pepper.

Beetroot on a barbie is a winner too. Wrap oiled whole fresh beetroot individually in a double layer of foil and chuck them in the embers of a hot barbecue for 45 minutes or so, and save them for lunch the next day.

BOAT PICNICS

Meandering along rivers and canals delivers up one perfect picnic spot after another, so yes, I'm going to talk about the boat picnic. Think food prepared with as little hassle as possible. Take advantage of ready prepared items, even eating out of the bowls you transport the food in so you don't have to carry lots of different things. As far as the rest goes, get a good picnic rug, some kitchen roll (which doubles as napkins), a few wet wipes, and you're good to go.

So, what to eat? Try a cold roast chicken, which you can either cook at home and bring on board or buy precooked from the supermarket. You prefer red meat? Fry a steak, slice it up, dress it with vinaigrette and serve it with a big, beautiful salad so everything stays juicy, and the flavours intensify.

Seek out a fresh crusty baguette, a large tomato and a whole salami, which you can slice as needed. And I always bring cheese to a picnic – hard cheese, such as a good strong cheddar or a nutty pecorino, but nothing too pongy or gooey which doesn't like the heat.

A vegetarian or vegan picnic can be equally satisfying. Think falafel wraps, frittatas (see page 105) and my favourite, classic Tuscan tomato and bread salad panzanella (see page 134). And always bring a bag of prewashed, crunchy lettuce to add freshness.

Remember that pasta salads can handle a bit of a battering during transit. And don't forget boiled new potatoes, including star of the show Jersey Royals, which when dressed with a little fresh mint and a perky vinaigrette are hard to beat.

For something sweet to finish, I take fruit such as cherries, and sometimes a shop-bought carrot cake, as it's easy to transport. If you want to cook something special then try my slab scone recipe with strawberries and clotted cream on page 139, which can be assembled and sliced into squares in situ.

Some advice about packing your picnic. Ice packs are essential to keep things cool if you're lucky enough to have a fridge and icebox on board. No fridge? Then an insulated cool box will still help keep temperatures down. Check your Tupperware boxes are tightly shut and wrap pungent items in compostable food wrap. Put dressings in old (clean) jam jars so they are ready to shake and drizzle.

Finally, what to drink? Think cool and thirst quenching. Rosé is generally considered the quintessential picnic wine but if you want to try something a bit different go for a light red, such as a Beaujolais, lightly chilled. On the white wine front, I gravitate towards lower-alcohol whites for alfresco lunches, such as German Riesling and Portuguese Vinho Verde. You also can't go wrong with fizz on a picnic – try Crémant from the Loire or Alsace. And bring water, of course. Everyone always forgets the water.

Spring

It's a time of celebration and renewal, the anticipation building from the first daffodil to the lighter, warmer days of late spring. With their crisp textures, vibrant colours and delicate flavours, spring vegetables offer the perfect opportunity to breathe new life into your cooking.

After a winter of hearty meals and hibernation, we're more than ready for something lighter and fresher tasting to eat. Spring delivers that in spades, from radishes to asparagus, herbs to rhubarb. Refreshing, regenerating, and detoxifying in equal measures. With their crisp textures, vibrant colours and delicate flavours, spring fruit and vegetables offer the perfect opportunity to reinvigorate your cooking.

WHAT'S IN SEASON?

Asparagus • Broad beans • Carrots • Cauliflower • Kale • Leeks • Lettuce • New potatoes • Purple sprouting broccoli (all year round, but best in the spring) • Radish • Rhubarb • Savoy and spring cabbages • Spinach • Spring onions • Watercress

SPRING FORAGING

My top 5: wild garlic, nettles, hawthorn leaves and blossom, wild fennel, dandelion

If I had to pick a favourite wild food, it would be **WILD GARLIC**. I've eaten it every which way, in soup and pesto, in pies and risottos, in scones and chicken Kiev (so good). It makes a good first stop on your foraging journey as it's easy to spot, just follow your nose. It's also a brilliant way to introduce children to foraging, pick it with them and then prepare it together for dinner. Every part of the wild garlic plant (also called ramsons) can be used to flavour food, but I like the young shoots best.

NETTLES must be the most ubiquitous of wild foods, growing on every verge, path and field. Super healthy, they are packed with iron and vitamins A and C, not to mention possibly lowering both blood pressure and heart rate. Just remember to wear gloves while picking them. And go for the young leaves – the greener bit at the top of the plant, or new plants just a few inches off the ground. Cooking or thoroughly drying nettles leaves and stems neutralises the stinging properties of the plant. Although the nettle is in leaf from March to November, the leaves are best picked when young and fresh, and can be handled much like spinach, washing and then cooking for a few minutes in the water that clings to the washed leaves, then serving with a twist of black pepper and some butter. I've even made nettle crackers for cheese using my sourdough starter.

HAWTHORN is a forager's favourite. One of the most common native British trees, it offers a bounty for the boat kitchen with its edible foliage, almond-scented blossoms, and pectin-rich berries later in the year (see Winter). As well as being widely available, it's easy to identify. In early spring, the young leaves add a nutty note to salads – you can use it in any recipe calling for spring greens. Pick the pretty blossoms on a warm sunny day

MAGICAL HAWTHORN

The hawthorn is steeped in folklore. The story goes that this tree is the threshold to the 'other world' and it's where fairies live. Also known as May blossom (as it often flowers on May Day), it happens to coincide with the pagan celebration of Beltane, which is dedicated to fairies. The five petals of the blossom form a pentagram, a magical sign known as the elven cross. And when the hawberries are ripe, they have a 5-pointed star at the base of the fruit. So, if you want to avoid being cursed, don't chop down a hawthorn.

for their funky almond scent and scatter over salads, or use them to make a cordial.

We've got the Romans to thank (once again) for **WILD FENNEL**. It's found all over the UK, along roadsides, fields and meadows. It's particularly delicious with eggs, fish, and in salads. Not to be confused with domesticated fennel, which boasts a large, edible white bulb, wild fennel yields feathery fronds come springtime, which can be picked, chopped and sprinkled liberally over whatever takes your fancy. I love it stuffed into the cavity of a fresh fish with a few thin slices of lemon. In summer, wild fennel's edible seeds and flowers add another dimension to your cooking. It's simple to identify, too, with its aniseed aroma.

Is there a wildflower more recognised than the **DANDELION**? It grows everywhere, all over the world, in pastures, meadows, waysides and lawns. It flowers profusely in April and the leaves can be found at any time of the year, except in the very coldest months. Good to look at, with an interesting bitter taste (more so as the summer draws on), among other health benefits, the dandelion flower has antioxidant properties and the root is rich in potassium. The young leaves are often eaten as a salad, especially in France, where they are known as *pissenlit* in markets, and can be cooked like spinach. The flowers can be fermented to make wine, while the roots can be used in salads, grated or chopped, or you can roast and grind the roots to make a coffee substitute. Little tip: wear gloves to avoid staining your hands while chopping.

Porridge with rhubarb and orange compote and toasted almonds

BREAKFAST

I could write a book about porridge. It boasts impressive health benefits – it's proven to lower cholesterol and protect against heart disease (if it's made the more austere way) and it boasts a low GI. And there are so many ways to make it. Purists will tell you that porridge should contain nothing more than oats, water and salt. But that's boring. Sure, you can keep it simple with some demerara sugar, or golden syrup, or Greek yogurt and a few berries, which I enjoy regularly. I buy whatever organic jumbo whole rolled oats I see on the shelves, but feel free to buy whatever oat type you prefer. This topping is one of my favourites – and one I look forward to every spring when the first rhubarb starts to appear. The toasting of the oats only takes a couple of minutes and gives a pleasingly nutty flavour to the porridge, though you can skip this bit if you can't be bothered. The rhubarb compote is also a winner with Greek yoghurt if you're ever stuck for a pud.

FOR 4 SERVINGS

For the rhubarb and orange compote:
500g rhubarb, cut into batons
1 orange, zested
2 oranges, juiced
50g caster sugar

For the porridge:
200g jumbo whole rolled oats
400 ml whole milk
800 ml water
Good pinch of salt
50g toasted flaked almonds (or hazelnuts also work well)

Place the rhubarb into a pan with the orange juice, zest and sugar. Cook gently until tender, about 8–10 minutes. Set aside to cool. Heat a non-stick frying pan over a medium-high heat and toast the oats until fragrant, stirring a couple of times (about two minutes). Put the oats in a saucepan along with the milk and water and bring slowly to the boil, stirring frequently. Turn the heat down to

low and simmer for 6-7 minutes, adding the salt halfway through and stirring regularly. Cover and let it sit for five minutes, then serve with the rhubarb compote and almonds, or toppings of your choice. I also like to add a slug of cold milk to the porridge just before serving.

Naan bread breakfast pizza

BREAKFAST

Hands up who's eaten leftover pizza for breakfast? Me! So, why not make a breakfast pizza? Not from scratch, obviously – that's way too much first thing in the morning. Though let's be honest, shop-bought pizza bases are generally not great. Shop-bought naan bread on the other hand is always pretty good. Do you see where I'm going with this? I'm not the first, promise – Nigella is a fan. Plus, the bacon naan at cult Indian restaurant chain Dishoom is still one of my all-time favourite breakfasts.

I've substituted bacon for Parma ham to speed things up (or you could use pancetta). Sometimes I even use one of those supermarket salad pots

of mozzarella pearls and slow-roasted tomatoes. Feel free to add sliced button mushrooms to complete 'The Full English on a Pizza' vibe.

FOR 2 PIZZAS

2 shop-bought naans
125g mozzarella
50g Gruyère cheese, grated
150g cherry tomatoes, halved (or jar of slow-roasted tomatoes – or both!)
4 slices of Parma ham, torn
2 medium free-range eggs
A few fresh thyme springs, or a pinch of dried thyme (optional)

Preheat the oven to 220°/fan 200°/gas 7. Lay the naans on a baking tray. Dot over the mozzarella and grated Gruyère, leaving a divot on each naan for the egg. Arrange the tomatoes on top, followed by the ham – always leaving a space for the eggs to keep them in one place. Crack the eggs and drop into each divot. Sprinkle over the thyme, if using. Bake for 12-15 minutes, until the white is set and the yolk still a little runny.

Wild garlic pesto

QUICK & LIGHT

Wild garlic is the foraged food I look forward to most, carpeting bosky woods with its pretty white flowers and deep green, glossy leaves. It pays to remember that when foraging for wild garlic – or any sort of foraging come to that – do apply the golden rule of taking only a third of the plant (see page 36).

I include wild garlic in everything from scrambled eggs to cheese scones, but pesto is at the top of my list. I chuck it over drained, cooled green beans, and stir it into pretty much any tinned beans I have in the cupboard, plus dollop it liberally on soups. I cut it with parsley to dial down the pungency, but your call. It will keep for at least a week in the fridge. Feel free to replace the pine nuts with any nut of your choice. This amount will fill an old (sterilised) jam jar and is enough to serve four added to pasta, allowing for 75g of pasta per person. I tend to make it at home and bring it on board, though sometimes I get out the pestle and mortar on board and make a chunkier version.

FOR 4 SERVINGS

70g wild garlic, washed
30g parsley (flat or curly)
60g pine nuts
60g parmesan
150 ml olive oil (I mix half extra virgin olive oil with half regular olive oil)
½ lemon, juiced (or to taste)
Salt and black pepper

Place all the ingredients into a food processor apart from the olive oil and blitz for a minute or two, then slowly pour in the olive oil until blended.

Asparagus and new potato frittata

QUICK & LIGHT

I make no excuse for revisiting asparagus frittata (there's a version with goat cheese in my *Boat Cookbook*). I love asparagus with eggs. I often use the spears as 'soldiers' for boiled eggs, and I'm a big fan of the Spanish tradition of adding baby asparagus to scrambled eggs. Combine it with new potatoes in an omelette, or in this more picnic-portable frittata (go one step further and use the king of new potatoes, the Jersey Royal) and it's a great way to showcase this magnificent spring vegetable. Serve it with a mixed salad to increase your vegetable intake further.

FOR 4 SERVINGS

200g new potatoes
100g asparagus tips, cut into 3 cm pieces
1 tbsp olive or rapeseed oil
1 onion, finely chopped
8 large free-range eggs, lightly beaten and seasoned
20g parmesan, grated
100g feta cheese

Halve the potatoes, then cook in boiling water for 8-10 minutes, before adding the asparagus and cooking for a further four minutes. Drain and set aside. Meanwhile, heat the oil over a medium heat in a non-stick frying pan, add the onion and soften for five minutes. Dice the potatoes and add with the asparagus to the onion to heat through for two minutes. Combine the eggs with the grated Parmesan and add to the potato, onion and asparagus. Crumble over the feta cheese and cook gently until the bottom is set, about 10 minutes. Meanwhile, heat the grill. Remove the pan from the hob and place under the grill until the egg is set and the top is lightly browned. Serve warm or at room temperature, cutting into slices to serve.

Chicken, spinach and lemon orzotto

BIGGER PLATES

Who wants to stand at the cooker stirring risotto for half an hour? Okay, me – I find it oddly relaxing. But not many of you, I suspect – especially when you've got guests on board. Say hello to orzotto. It's basically risotto made with orzo pasta, which is less needy than rice. Once the orzo goes in the pan, you give it a good stir, add the stock and the meat, and set the timer for 12 minutes, stirring once halfway through, and ta-dah! A quick, classic one pot dish, lifted by the addition of lemon.

FOR 4 SERVINGS

2 tbsp olive oil
4 skinless, boneless chicken thighs, chopped into bite-size chunks
Salt and freshly ground black pepper
1 large onion, finely chopped
2 garlic cloves, chopped
300g dried orzo
1 tbsp Dijon mustard
750 ml chicken stock
200g baby spinach
1 lemon, zested and juiced
2 tbsp crème fraîche or sour cream
3 tbsp parmesan cheese, grated

Heat 1 tbsp of the olive oil in a non-stick sauté pan or shallow casserole and brown the chicken with a little salt over a high heat, about five minutes. Transfer to a plate. Lower the heat, add the other tablespoon of olive oil and the onion, and cook for five minutes until soft, then add the garlic and cook for another minute. Add the orzo and the mustard, stirring until coated. Add the stock and the browned chicken, bring to the boil, and season. Reduce the heat and simmer uncovered for 12 minutes, stirring occasionally. Add the spinach and stir until wilted. Add the lemon juice and zest, crème fraîche or sour cream and, parmesan and stir, then serve.

Prawn and chickpea stew with leeks and lemon

BIGGER PLATES

This fresh, vibrant seafood stew is so much more than a sum of its parts. It's simple to make yet delivers layers of flavour – and is particularly satisfying when you've got some fresh crusty bread to mop it all up. All it needs is a bright, zesty white to wash it all down – think Spanish Verdejo, or Italian Verdicchio, and you'll have this on repeat. Best are those Bold Bean Co chickpeas you get in jars in supermarkets, but regular tinned chickpeas are just fine, or indeed any tinned white beans. To keep costs down, you can use supermarket bags of frozen raw king prawns, letting them defrost as you cart them back to the boat if you're cooking them that day.

FOR 4 SERVINGS

1 lemon, zested and juiced
1 tsp smoked (mild) paprika
2 garlic cloves, crushed
Salt and black pepper
450g raw jumbo king prawns, peeled
50g unsalted butter
2 large leeks, trimmed, finely sliced and washed
2 x 400g chickpeas, rinsed and drained
1 litre vegetable stock
A handful of flat leaf parsley, finely chopped
Crusty bread for mopping up the juices (optional)

Combine zest, paprika, garlic, ½ tsp salt and ½ tsp pepper in a bowl. Add the prawns and toss to coat. In a large pot, melt the butter over a medium heat, add the prawns and cook, stirring occasionally, until pink, about 2-3 minutes. Using a slotted spoon, transfer the prawns to a plate and set aside. Add the leeks, season with salt and pepper and cook with a lid on over medium heat until the leeks are soft, about eight minutes. Add the drained chickpeas and the stock and bring to a simmer, then cook for five minutes. Stir in the prawns and any juices from the plate, plus the parsley and the lemon juice. Season, then serve with the bread, if using.

Alphonso mango with coconut and mascarpone

SWEET TREATS

Spring brings many delicious treats, but when it comes to fruit, the Alphonso mango is up there. Come April, Asian greengrocers proudly display them (they don't grow here in the UK), with their sunshine yellow, smooth buttery flesh, and heady peachy, citrus and honey flavours. They have just a short season, from April until late June, so cherish them. In India, where there is a national obsession with Alphonso mangoes, it's customary to give boxes of them to friends and family to share the love. I buy them online from Natoora when I'm not within grasp of an Asian greengrocer. And if you can't lay your hands on Alphonso mangoes, then any ripe mangoes will do. I use Rachel's Greek-style coconut yoghurt for this recipe, which is widely available.

FOR 4 SERVINGS

2 ripe Alphonso mangoes, or 400g ready-diced mango
1 lime, zested and juiced
125g coconut yoghurt
125g mascarpone
1 tbsp honey
1 tbsp of peeled coconut, toasted

Peel and cube the mangoes and squeeze over the lime juice. Put half of the mango mixture into a pestle and mortar and pound for a minute to make a rough purée. Combine the yoghurt with the mascarpone and honey and divide between four glasses. Top with the cubed mango, and garnish with the lime zest and toasted coconut.

Sea salt chocolate brioche toasties

SWEET TREATS

I love how brioche is both light and rich. Add chocolate and it's a wonderfully indulgent treat with your morning coffee or afternoon tea. Brioche loaves are now available in most supermarkets. For the chocolate, make sure you buy it minimum 70% dark. The salt just brings out the flavours and provides a satisfying foil for the sweetness.

FOR 2 SERVINGS

4 thin slices of brioche
100g dark chocolate (minimum 70% cocoa solids)
Pinch of Maldon salt

Heat a ridged griddle pan or non-stick frying pan. Lay two of the brioche slices flat on a work surface and grate over the chocolate, spreading it out almost to the edge. Sprinkle over a pinch of crumbled Maldon salt on each. Cover with the other slices. Toast on the griddle until the underside is done before carefully turning so that the chocolate doesn't escape, and toast the other side. It's ready when the chocolate has melted and the brioche is golden.

Peanut butter biscuits

SWEET TREATS

Peanut butter is my guilty pleasure. And smooth peanut butter at that, ideally spread on hot toasted white sliced bread. So this is for anyone out there who feels the same. These biscuits are mostly just peanut butter held together with sugar and egg, and equally simple to make, with no major kitchen gadgets required. You can add chocolate if you fancy – around 30g, chopped, though keep it dark at 70% cocoa solids or above. You can use crunchy peanut butter too, if you prefer. I always choose the healthy kind for this recipe: the one without any added sugar as there is sugar enough in the dough. The biscuits are best enjoyed while slightly warm. What you want is a crunchy edge and a chewy middle, so you might need to play around with the cooking times depending on how hot your boat oven runs. Just try not to scoff them all at once.

MAKES 10-12 COOKIES

1 medium egg
100g caster sugar
200g smooth peanut butter
1 tsp ground cinnamon
Pinch of flaky sea salt

Preheat the oven to 190°/fan 170°/gas 5. In a large bowl, beat the egg and the sugar, then add the peanut butter and ground cinnamon, beating again until it thickens (the chocolate goes in now if you want it). Shape the dough into walnut-sized balls and place on a baking tray lined with baking parchment. Press each ball down with a fork, top each with a smidge of crumbled Maldon salt and bake for 12-14 minutes. Leave to sit for a few minutes before cooling on a wire rack. The biscuits will keep for a few days in an airtight container.

TIPPLE TIME

WHAT TO DRINK

The daffodils are blooming, the scent of newly cut grass fills the air, the weather is warming – hello spring, my favourite season! I love the sense of anticipation as the clocks jump forward and the nights get longer. Covers are whipped off barbecues, and back gardens and boat decks hum with gentle chatter again as we take to the outdoors, albeit wrapped in a sweater or two. And as we shrug off the colder months, we move from hearty reds, richer whites, and creamier cocktails to something brighter and breezier in our glass.

To celebrate this most-looked-forward-to season, think fizz, from non-vintage Champagne to Crémant, Cava and Prosecco – and English sparkling wine, of course, with its crisp acidity and hedgerow blossom aromas, a truly fitting spring drink. The season's bright, new vegetables, from asparagus to purple sprouting broccoli, need equally bright wines – think herbaceous Vermentinos, and minerally Pinot Grigios.

For that first alfresco spring lunch of grilled fish and vegetables, look at Albariño and Picpoul de Pinet, which will match those briny seafood flavours a treat. For roast chicken and salad, try a light unoaked Chardonnay or an Assyrtiko from Santorini in Greece, now starring on a supermarket shelf near you. For crisp green salads spikily dressed, look to Rieslings and Austrian Grüner Veltliner, or crisp, mineral-driven Sauvignon Blancs.

You're a red wine drinker? Ditch the heavy reds and look to lighter, crunchier red wine styles that you can happily glug without food. France does this type of red wine better than anyone else – they call it *vin de soif* (thirst quencher), from Cabernet Franc-based Loire reds to simple Beaujolais Villages.

Italy, too, has a wealth of lighter style reds to try, from Bardolino and Dolcetto, to Frappato and Valpolicella; and try Spanish wines made with grape varieties such as Mencia or a young Tempranillo. Further afield, also seek out cooler climate Pinot Noirs for spring drinking. Do consider chilling this style of red wine, too, on warmer days – not fridge cold, just a 20 minute or so blast in the fridge (or the river, attached tightly with rope) to take the temperature down a notch or two.

For spring-ready beers, seek out refreshing brews to enjoy in the watery sunlight, singling out those boasting floral notes, citrussy hops, or hints of flowers. And yes, beer producers do brew with the seasons in mind, numerous craft breweries especially, offering a variety of styles to suit the weather. Look

at styles such as wheat beers, pale ales, sour beers, and fruit-infused beers for spring beer drinking.

April showers bring May flowers so why not reflect that in your cocktails too? Seek out fresh, light flavours in cocktail recipes, from floral drinks to frothy egg white concoctions, plus there is plenty of inspiration out there to make the most of the season's best produce, from rhubarb to blood oranges. Classic spring cocktail favourites include Gin Fizz, Tom Collins, and the Lemon Drop, though if I had to pick a favourite it would be another lemony classic, French 75 – gin, lemon juice, and simple syrup topped up with Champagne. In fact, lemon is a common theme in my drinks come spring.

THREE GREAT SPRING COCKTAILS

Blood orange and vermouth cobbler

This was created by head mixologist Edoardo Sandri at the Atrium bar in the Four Seasons Hotel in Florence. A clever riff on a classic sherry cobbler, it's perfect for spring when you're looking for a lift as the days lengthen. It also takes advantage of the end of blood orange season, which runs through to

May. If you can't get hold of blood oranges, just use regular oranges. To make crushed ice on board, place cubes into the middle of a clean tea towel and fold in half, then bash with a rolling pin and shake the crushed ice into a bowl, ready for use.

1 blood orange
1 tsp sugar
80 ml white vermouth (I like Noilly Prat)
2 sprigs mint
10 ml Amaretto (I use Disaronno)

Cut a twist of peel off your orange, setting aside to use as a garnish. Then cut the orange in half horizontally. Cut half into thick round slices and cut each into quarters, placing into the bottom of a tall glass. Sprinkle over the sugar and leave to macerate for a moment. Add the white vermouth to the glass, rub one of the sprigs of mint in your hand to release the scent and add to the glass. Fill the glass to the top with crushed ice. Pour the amaretto over the top of the ice and garnish with the twist of orange peel and the remaining sprig of mint.

Rhubarb and lemon fizz

The colour and taste of the first of the season's rhubarb fills me with excitement. This is a great way to celebrate this beautiful vegetable. And yes, it is a vegetable, not a fruit, part of the Polygonaceae family, which includes buckwheat. It's also a truly stunning addition to any cocktail.

2 shots vodka
1 shot rhubarb simple syrup (see overleaf)
1 lemon
Soda or sparkling water
Twist of lemon peel to garnish

Cut a twist of peel off your lemon, setting aside to use as a garnish. Then squeeze the lemon until you have one shot of lemon juice. Add the vodka, rhubarb simple syrup, and lemon juice to a cocktail shaker and fill with ice. Shake vigorously and strain into a glass filled with fresh ice. Top with the soda water or sparkling water and garnish with the twist of lemon peel.

HOW TO MAKE RHUBARB SIMPLE SYRUP

The colour alone should win you over. It adds a pinch of tartness and a beautiful hue to any drink. I make simple syrup with less sugar than most as I don't like sweet drinks. Most simple syrups are equal parts sugar to water, but mine is nearly two parts water to one part sugar.

250 ml water
125g rhubarb, chopped into 2 cm pieces
100g sugar

Combine all the ingredients in a saucepan and bring to the boil. Reduce the heat and simmer for 20-25 minutes until the rhubarb falls apart and the mixture has thickened. Take off the heat and allow to cool completely. Strain into a glass jar through a fine mesh strainer, gently pressing the fruit to extract as much gorgeous juice as possible. It will keep in the fridge for several weeks.

Rooibos and honey mocktail

I drink a lot of rooibos tea after discovering it many years ago during my first visit to South Africa, where it's widely glugged, chilled in the summer months and sipped hot the rest of the year, with or without milk. It's naturally caffeine-free and it's known as an anti-inflammatory and an antioxidant. A cultivated crop since the 1930s, it's now widely available all over the world where it's also known as redbush tea. It makes a good mocktail too.

1 rooibos teabag
2 tbsp honey
50 ml water
Tonic water
Slice of dehydrated lemon to garnish (optional)

Infuse the teabag and honey in 50 ml hot water for 10 minutes. Strain into an ice-filled glass, top with tonic and garnish, if you like, with the lemon wheel.

SPRING BEER SNACK

Prosciutto, crisps, Guindillas peppers

I got this pile-'em-up presentation idea from a brilliant Basque-themed restaurant in London called Ibai. Chef Richard Foster and his brigade make their own potato crisps, plating them up with layers of sweet, nutty Le Noir de Bigorre (an AOP ham made with ancient breed of pig from the Pyrénées), topping the lot with a handful of sharp, smoky, Piparras peppers. Instead, I've used prosciutto because it's readily available, but you could mail-order in a stash of Spanish ham from Brindisa.com. And if you can't find Piparras peppers, try spicy, tangy Guindilla peppers, also from the Basque country, found in most supermarkets.

FOR 4 SERVINGS

125g potato crisps
90g prosciutto, torn into large strips
Handful of Guindilla peppers, drained and patted dry with kitchen paper

In a large bowl, layer the crisps with the ham and top with the drained peppers.

Summer

It's about living life on the veg, is summer. A season of some truly iconic crops, a total joy from start to finish, when everything is super fresh. Eat your fill when these wonderful ingredients are at their best.

Is there a more thrilling time of the year to eat seasonally than the summer? Everything is at its jewel-like best, fresh and fragrant, and from earth to plate often in minutes. It's the season for barbecues and picnics, and for long meals alfresco, with minimal cooking and maximum flavours. Even a glut of courgettes won't get you down, as you will find ever more inventive ways to serve them. In short, we're spoiled for choice when everything is bursting with vitamins and minerals, promising optimum flavour and nutritional value.

WHAT'S IN SEASON?

Assorted herbs • Aubergines • Baby carrots • Beetroot • Blackberries • Blueberries • Broad beans • Celery • Chard • Cherries • Chillies • Courgettes • Cucumber • Currants • Fennel • Figs • French beans • Globe artichoke • Gooseberries • Greengages • Lettuces and other leaves • Loganberries • Mangetout • New potatoes/potatoes • Peas • Peppers • Plums • Radish • Raspberries • Rocket • Runner beans • Shallots • Spinach • Spring cabbage • Spring onions • Strawberries • Summer squash • Sweetcorn • Tomatoes • Watercress

SUMMER FORAGING

My top 5: elderflowers, sorrel, blackberries, crab apples, damsons

Once you've tasted homemade **ELDERFLOWER** cordial, you'll never buy another bottle. Its sweet, fragrant, soporific scent is summer in a glass. June is generally the best time to pick elderflowers, when they are at their freshest and beautiful best. You can add them to cakes and biscuits, but nothing beats elderflower cordial. Serve with water and a slice of lemon and it's the quintessential summer thirst quencher, but it also adds an intriguing twist to cocktails (see page 146). The last of the great tree flowers of the year, the elderflower also boasts impressive health benefits, with both antiseptic and anti-inflammatory properties. Little tip: pick the flowers on a warm, dry day.

The French love **SORREL**, but we don't use enough of it these days. Historically, it was typically used in a green sauce served with cold meats, pounded by pestle and mortar, and mixed with vinegar and sugar in the manner of mint sauce. A perennial herb, it grows throughout the British Isles, on grassland and in open woodland, flowering from May to August. I love the flavour of sorrel – it's tart, lemony and grassy, and a great substitute for lemon or lime. It's particularly fetching when cooked with potato in a soup, but it's also a favourite in a hollandaise-style sauce (I use a recipe from the late, great Jane Grigson), served with poached salmon or trout.

Number two on my all-time favourite wild food list is **BLACKBERRIES**. The anticipation is huge, as summer rolls on and the brambles slowly ripen. Come August, everyone is out picking. I always keep these jewel-like fruits in the freezer at home to pimp up an apple crumble or pancakes (see

page 130), and to make jam. Or I throw a handful into my muffin or cake mix to add a tart juiciness. With their high antioxidant content, they're more than just a pretty little berry and thankfully grow in abundance around the country. Little tip: wash well with cold water and leave to soak with a little salt to kill off any bugs.

I ignored the **CRAB APPLE** for years. Impossibly sour (it's the origin of the expression 'crabby', meaning a grumpy person), I wasn't drawn to it. But as more chefs around the country practise being hyper-local and find intriguing ways to get acidity into a dish without using imported lemons, the crab apple has moved into the spotlight. In fact, as a substitute for lemon juice it's even better than sorrel – you can make verjuice out of crab apples (see below), which can be used in place of lemon, or even wine, come to that. It also makes an exquisite, jewel-like jelly, which is fabulous on scones and works well as a fruit leather. The crab apple tree thrives all over the British Isles, the fruit ripening in late summer.

> **MAKING YOUR OWN VERJUICE WITH CRAB APPLES**
> This is the best substitute for lemons I've ever come across. Fill a basket with crab apples, then clean and sort the fruit. Pop into a juicer. Then pour the liquid into a sterilised jar, removing any excessive foam. Cover with a cloth and leave in a dark, cool place to ferment for a month. Then strain out the gunk that has gathered at the bottom of the jar, washing the jar out, before pouring the filtered juice back in. Keep it in the fridge and use in place of lemon or wine, adding to sauces, risottos and vinaigrettes. It's also great used in a marinade for chicken.

DAMSONS are often associated with old lock cottages, so its inclusion is a must. Brought to the UK by the Romans, they grow wild in hedges, on heathlands, and in parks. When fully ripe – usually around August - they provide intense yellow or dark purple fruit with a single stone that can be eaten on the spot. If it's too acidic, then make into jam, jellies, liqueurs, ketchups and chutney. The trees never grow too tall either, so the fruit is pretty easy to reach, growing in small bunches like cherries.

Boat granola

BREAKFAST

Now here's the thing – I only toast the oats in the oven for my granola. I like to keep the seeds and nuts raw, stirring them in later. It's for the flavour, certainly, but also to maintain max nutritional value. The granola is great used as a healthy topping on fruit, whether in compote or salad form. It keeps well, too, in an air tight container for up to a month. I like to serve it with yoghurt or kefir, fresh berries, or chopped apples or pears.

MAKES A ONE LITRE JAR

3 tbsp coconut oil (or vegetable or olive oil)
1 tbsp honey
250g rolled jumbo oats
Pinch of sea salt
100g mixed whole nuts, such as walnuts, pecans, almonds, and hazelnuts, roughly chopped
100g mixed dried fruit, such as sultanas, chopped apricots and prunes and goji berries
50g mixed seeds, such as pumpkin and sunflower

Preheat the oven to 180°/fan 160°/gas 4. In a small saucepan, warm the oil and honey, stirring well until combined. Add the oats into a large bowl together with the salt. Pour over the warm sweetened oil, mixing well so the oats are thoroughly coated. Spread the oat mix on a baking tray lined with baking paper and cook for 20-25 minutes or until golden, stirring halfway through. Leave to cool completely. Once cool, mix in the chopped nuts, the dried fruit and seeds, and transfer to an airtight container or jar.

Banana pancakes with wild blackberries

BREAKFAST

I've loved banana pancakes ever since travelling around Southeast Asia post-uni. These are more of an American style pancake. I always use the American cup measurement system, which I've included, for this, because it's so easy, especially on board. Pairing the pancakes with intense, sun-ripened wild blackberries is even more of a treat, though failing that, use whatever berries you can lay your hands on or just add sliced banana.
In America, it's traditional to serve pancakes with maple syrup, but I love honey, especially as Liz produces her own.

FOR 4 SERVINGS, 8 PANCAKES

1 cup (135g) plain flour
1 tsp baking powder

½ tsp cinnamon

¼ tsp salt

¾ cup mashed ripe banana (about 2 medium bananas)

1 large egg

¾ cup milk (185 ml)

Sunflower oil cooking spray

Wild blackberries for serving

Honey for serving

Measure out the flour, baking powder, cinnamon and salt in a bowl and whisk together. In a large bowl, mash the banana with a fork, add the egg and whisk together until blended. Whisk in the milk until combined, then whisk in the dry ingredients – the batter will be slightly lumpy, but that's fine. Heat a large non-stick frying pan over a medium heat and spritz with cooking spray (I've found this is the easiest way to cook pancakes). Fill the ⅓ cup measure (or a ladle) and pour circles of batter into your pan – I can get two in my pan at a time. Cook until puffed up and golden underneath, about two minutes, then flip and cook on the other side until risen and cooked through – another two minutes. Keep them warm, while you repeat the exercise. Serve with the blackberries or a topping of your choice, and drizzle with honey.

Hot smoked salmon salad with watercress and fennel

QUICK & LIGHT

This is such an easy summer salad – it looks great, too, with its flecks of pink fish against the vibrant green of the leaves. I sometimes serve it with a vinegary potato salad on the side if I want to beef the meal up a bit, or just some crusty bread to scoop up the juices.

FOR 4 SERVINGS

1 head romaine lettuce, leaves torn and washed

80g watercress, washed

300g hot smoked salmon fillets, skin removed, separated into chunky flakes

½ fennel bulb, thinly sliced (on a mandolin if you have one, see Kit page 18)

2 tbsp salted capers, rinsed

4 spring onions, trimmed and sliced
10g dill, chopped
4 tbsp olive or rapeseed oil
1 lemon
Black pepper

Combine the first seven ingredients in a large bowl. Drizzle over the oil, squeeze over the lemon, add a twist or three of pepper, serve.

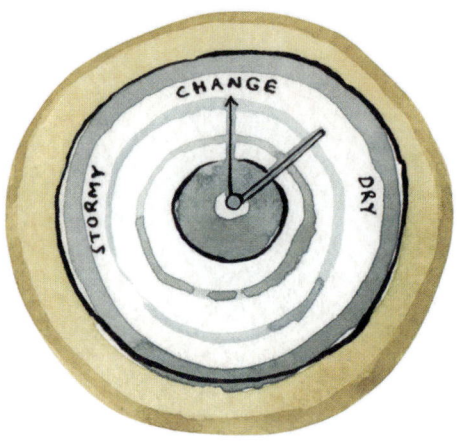

Puy lentils with roasted sweet peppers, tomatoes and feta

QUICK & LIGHT

I always have a stash of precooked lentils in my cupboard, whether on board or at home. There's just so much you can do with them. Ditto a jar of roasted peppers. This dish is one of my favourite ways to dress them up. Pulled together in minutes, it makes for a substantial summer supper. You can also serve it as a side for fish or meat, then I just omit the feta.

FOR 4 SERVINGS

250g precooked Puy lentils
200g cherry tomatoes, halved
200g jar of roasted red peppers, drained and roughly chopped
1 small red onion, finely diced
Handful of basil leaves, chopped
Handful of fresh mint, chopped
25 ml extra virgin olive oil
25 ml balsamic vinegar
100g feta
Maldon sea salt and black pepper

In a large bowl, combine all the ingredients, crumbling in the feta, stirring gently until well mixed. Season to taste and serve at room temperature.

Panzanella

QUICK & LIGHT

This Italian bread salad is one of my all-time favourites. The flavours mingle to produce one soft, juicy, aromatic jumble. It's also a brilliant way to use up stale bread. It's important to leave it to sit for a bit to soak up all those flavours. The tomato juices are key, so don't waste them while chopping. It's good enough as a stand-alone dish, but it goes particularly well with barbecued meats (see the chop recipes on pages 138 and 160).

FOR 4 SERVINGS

250g day-old good quality bread (such as sourdough or ciabatta)
Half a 480g jar roasted red peppers, drained, and roughly chopped
600g mixed tomatoes, roughly chopped
1 small red onion, halved and thinly sliced
1 tbsp capers
100g good quality black (or green) olives, pitted and roughly chopped
8 anchovies, or to taste, finely chopped (optional)
A good pinch of salt
1 garlic clove, crushed

3 tbsp red wine vinegar

6 tbsp extra virgin olive oil

Freshly ground black pepper

A small bunch of basil, leaves picked and torn

Tear the bread into 2-3 cm pieces and leave on a tray to dry out for 30 minutes or so. Place the peppers in a large bowl with the tomatoes, the onion, the capers, olives, and anchovies, mixing well to combine. Add the bread and the salt and let it all sit together at room temperature for about 30 minutes for the juices to sink in and the flavours to combine. In a small bowl, mix the garlic, vinegar, and oil, seasoning with salt and black pepper. Just before serving, add the dressing and the torn basil leaves and give it all a good stir.

Chef Recipe – Vanessa Marx

Roasted chalk stream trout with fennel, herbs and citrus

BIGGER PLATES

Diners gaze out over riverboats chugging along the Thames at Richmond. They nibble on South African-born chef Vanessa Marx's fiercely seasonal menu at Bingham Riverhouse, a characterful members' club with 14 stylish bedrooms. It's all about sustainability here, a passion project for owners the Trinder family and Marx herself, with so many eco-friendly measures in place that it was the first independent hotel in the UK to receive a coveted Certified B Corporation status. The restaurant is open to all, and it's here that Marx dreams up her delicious plates of food using carefully chosen, locally sourced ingredients. This recipe has been kindly adapted by Vanessa Marx for the book. Chalk stream trout raised in nearby Hampshire is a regular on her menu. It became a firm favourite during the summer months, bursting with freshness, zest and crunch. It might become yours, too.

FOR 4 SERVINGS

1 orange

1 lemon

1 large fennel bulb, thinly sliced
20g chives, chopped
20g dill, chopped
4 medium chalk stream trout fillet portions, or use salmon fillets
50g butter, softened
50 ml olive oil
Maldon sea salt and black pepper

Preheat the oven to 200°. Line a baking dish with greaseproof paper to make a nest. Slice half of the orange, and half of the lemon. Add the sliced citrus into the dish, along with the sliced fennel bulb, chives and dill. Roughly mix to disperse the herbs. Place the trout portions on top of the fennel and citrus mix, with the skin side up. Squeeze over the juice from the remaining half of

the orange and lemon. Rub the outside of the fish with butter and drizzle with olive oil. Season with salt and pepper. Place the fish in the oven to roast for 15 minutes. Remove from the oven and drizzle with a little extra lemon juice and olive oil to serve.

Rosemary, garlic and honey pork chops

BIGGER PLATES

I make this marinade a lot for meat. It works particularly well with pork. Seek out chops with the bone in, as the taste is far superior, and they're usually juicier. This is perfect for the barbecue, but you can cook the chops on a griddle pan on the hob if you prefer. Serve with a side of your choice but I like it with the panzanella (see page 134). And remember to let the meat rest for five minutes before eating for extra tenderness.

FOR 4 SERVINGS

2 large bone-in pork chops

For the marinade:

2½ tbsp red wine vinegar

1 tbsp olive or rapeseed oil

1 tbsp honey

3 garlic cloves, minced

1 heaped tsp of finely chopped rosemary

½ tsp dried thyme

Pinch of Maldon salt and black pepper

Mix the marinade ingredients together in a bowl (I shake them up in an old jam jar) until the salt dissolves. Add to the chops and rub in to coat, then leave it to sit for two hours in a cool place. Meanwhile, prepare the barbecue. Brush the grill with oil if needed and heat to medium (or heat ½ tbsp of oil in a frying pan on medium to high heat). Cook the chops to your taste. Transfer to a plate and cover loosely with foil while the meat rests, then slice to serve.

Big scone, strawberries, and clotted cream

SWEET TREATS

Scones, strawberries and cream for pudding? Why not. Food blogger and chef 'The Boy Who Bakes' Edd Kimber introduced his slab scone a few years back, which got his TikTok followers all excited. This is my smaller, boat version, which is perfect for a picnic too, assembled once at your perfect spot. Scone dough is the easiest thing to throw together. And where to start with clotted cream? Divine, in a word. I think the mint finishes it, as both a colour contrast and a taste experience. Scoff the lot the same day, as it really doesn't keep. I've served it as a birthday cake before now, candles and all. I use the same useful square tin that I use for my chocolate tiffin (see page 195).

FOR 4 SERVINGS

250g self-raising flour, plus extra for dusting

1 level tsp baking powder

25g caster sugar

Pinch of salt

75g unsalted butter, diced
50 ml whole milk
2 eggs, lightly beaten
1 tbsp demerara sugar

For the topping:
200g fresh strawberries, hulled and halved (or quartered if large)
1 tbsp caster sugar
200g clotted cream
Handful of fresh mint leaves, chopped (optional)

Preheat the oven to 190°/fan 170°/gas 5. Line the base of a 23cm square loose-bottomed tin with baking parchment. In a large bowl, mix the flour, baking powder, sugar and salt together, then rub in the unsalted butter until it resembles coarse breadcrumbs. Make a well in the middle and pour in the milk and the beaten eggs, stirring to form a soft dough (it will be stickier than normal scone dough). Tip the dough straight into the tin and carefully spread the dough evenly to fit the tin. Sprinkle the dough with the demerara sugar, then bake for 20 minutes, or until golden brown. Set aside to cool completely in the tin.

Meanwhile, prepare the topping. Place the strawberries into a large bowl and sprinkle over the sugar, stirring together briefly. Leave to macerate for 40 minutes, or until the sugar has dissolved. Once cool, remove the scone from the tin and place it onto a plate. Spread the clotted cream all over the scone and then top with the macerated strawberries, drizzling with any syrup left in the bottom of the bowl. Add the mint garnish, if using. Cut into portions and serve immediately.

Grilled peaches with honey mascarpone and rosemary

SWEET TREATS

Barbecuing stone fruits is a revelation. The act of grilling transforms even the blandest of supermarket stone fruit into something intensely sweet and nutty. In fact, you don't want anything too ripe, or the fruit won't stand up to the heat. You're looking for firm fruit with a bit of give. The honey mascarpone just rounds things off nicely.

FOR 4 SERVINGS

4 large, mostly ripe peaches
2 tbsp olive oil
4 tsp brown sugar
250g mascarpone
1 tsp vanilla extract
4 tbsp runny honey
2 rosemary sprigs, chopped

Wash the peaches and slice lengthways to remove the stone. Brush each peach with olive oil on the cut side then sprinkle with brown sugar. Heat a grill pan or barbecue, then add peaches skin side up and cook until soft and slightly charred.

    Meanwhile, put the mascarpone, vanilla extract, and two tablespoons of the honey in a bowl and whisk together until light and combined. When the peaches are cooked, serve each half with a spoonful of honey mascarpone, a drizzle of the remaining honey, and a scattering of freshly chopped rosemary.

TIPPLE TIME

WHAT TO DRINK

Is there anything more bucolic than boozing by the water on a warm summer's day? But what to drink?

First, wine. Drinking wine in the summer months requires a different approach. Lighter, fresher wines take over from more weighty glugs – not only are they more thirst-quenching, but they also suit summer cooking better too.

Some wine styles work better than others. Think light, fresh whites. Look to varieties such as Sauvignon Blanc, Riesling, Verdejo and Albariño, Picpoul de Pinet and Muscadet, and yes, old favourite Pinot Grigio.

English white wines made from the Bacchus grape are quite simply summer in a glass, with notes of freshly mown grass and hedgerow aromas. English fizz, too, offers pleasingly bone-dry fruit with a fresh acidity and just the right amount of biscuity, yeasty flavours to lift the palate on a warm day.

German Riesling is one of the best value white wines in the world and you'll be hard pushed to find a food that doesn't go with it, plus it has a generally lower abv, which works better in warmer weather. You should also look at the limey, racy Rieslings from other spots around the world, most notably from Australia's Clare and Eden Valleys.

And not forgetting good value southern Italian whites, particularly those made from indigenous grapes such as Fiano, Grecanico and Grillo, offering a refreshing, zippy alternative to mainstream varieties.

Rosés rule in the summer and the style's popularity shows no signs of abating. Premium offerings are now de rigueur, and a rising number are made right here in the UK, such as Nanette's Rosé, from Balfour Winery in Kent. Little tip: always buy the most recent vintage; it will taste fresher, as most rosés don't age well.

Finally, red wine. Yes, some reds do make for great summer drinking. Go for simple, fruit-forward wines in a style that works when served slightly chilled, such as Beaujolais and other wines made from the Gamay grape from areas such as the Loire Valley, Oregon, and South Africa. You can also look at lighter styles of Pinot Noir, some Loire Valley Cabernet Francs, and Italian Frappato, Dolcetto et al. If in doubt, seek out those red wines on the shelves that are paler in colour as it suggests lighter fruit extraction and lower tannin levels.

A cold beer on a warm summer's day is one of life's little joys and there is now a dizzying array of summer beers to choose from, made within these very shores. My current favourite is award-winning South Wales brewery Tiny Rebel's Easy Livin' Pale Ale. From tart sour beers to tropical IPAs and fruit beers, there is a vast variety of summer brews just waiting to be explored, especially at the many riverside and canalside pubs around the country.

On the cocktail front, nothing says summer like a crisp G&T, and there are plenty of homegrown examples to explore. And here's a fun, less boozy twist: try sherry and tonic (fino, please). Sales of tequila have soared in recent years, and a Margarita is still one of my top three cocktails. But you could also try instead a Paloma, which has been gaining ground – tequila, grapefruit soda, and lime juice added to an ice-filled glass.

Spritzes abound on drinks lists around the country come summer and they make a good lower-alcohol choice. While Aperol Spritz still rules, try a Vermouth and tonic instead, preferably one of the growing numbers of excellent artisanal examples from both here and abroad.

Savoury drinks, too, are rising in popularity, as people seek out lower-sugar options with ingredients ranging from seaweed to turmeric. Spiced rums meanwhile are continuing to boom, with some made right here in the UK, while sparkling sake and sake-based cocktails are a relatively new kid on the block here that deserve some attention.

For all you NoLo drinkers out there, it has never been so good. The choice is now seemingly endless – from impressive zero-alcohol gin alternatives and decent sparkling rosés, to apple cider-based zero-alcohol cordials that satisfy even the snootiest of wine buffs. I'm looking at you, Matthew Jukes, and your excellent Jukes Cordialities (available online).

THREE GREAT SUMMER COCKTAILS

Ellenene fizz

This book – and the boat – needs a signature cocktail and this is it. For me, the ultimate combination is when the drink is finished with Chapel Down's A Touch of Sparkle (from Waitrose or online) made by the UK's largest wine producer, based in Kent. Its aromas of pineapple, grapefruit and elderflower, and its crisp, fresh palate sing perfectly with the gin, lemon and cordial. But it's also pretty good with any decent sparkling wine, English or

otherwise. Come late May, when the elderflower starts to bloom, add a sprig of blossom to the garnish.

FOR 1 GLASS

25 ml gin
1 tbsp lemon juice
1 tbsp elderflower cordial
Sparkling wine, for topping up
Twist of lemon peel to garnish

Combine the first three ingredients in a tall glass filled with ice. Top with the sparkling wine and stir. Decorate with a lemon twist.

Peach Bellini

The Bellini is a cocktail with an impressive provenance. It was first poured in the summer of 1948 by Giuseppe Cipriani, founder and barman of the legendary Harry's Bar in Venice. You'll find the entrance tucked away in a side street just off the Grand Canal. Once frequented by the likes of Humphrey Bogart and Ernest Hemingway, these days it's mainly for tourists (me included), who flock to sip this beautifully hued cocktail inspired by the region's fragrant white peaches and its famous sparkling wine, Prosecco. Cipriani's recipe calls for two parts Prosecco to one part fresh peach purée, gently stirred and served in chilled flutes.

FOR 8 GLASSES

4 ripe peaches
25 ml sugar syrup, or to taste
1 bottle of Prosecco

Blanch the ripe peaches in boiling water for two minutes. Plunge them into cold water so the skin can be removed easily. Peel the skin off and mash with your hands in a bowl, and then place the mashed peach and juice in a sieve, pushing down with a wooden spoon to extract as much pulp as possible. Stir in the sugar syrup. Then stir in 50 ml Prosecco so the pulp is absorbed and does not separate. Pour into chilled straight-sided glasses then top up with chilled Prosecco. Give it all a stir with a long spoon before serving.

Picnic citrus sling

Not boozing? Well, you won't miss out with this zesty concoction, which will hydrate as well as boosting your daily intake of vitamins and minerals.

FOR 4 GLASSES

400 ml cloudy apple juice
400 ml grapefruit juice (pink, preferably)
60 ml freshly squeezed lime juice
60 ml freshly squeezed lemon juice
75 ml sugar syrup
Lemon or lime wedges to garnish

Mix up all the ingredients and pour into a bottle to store in the fridge for a few hours, or even a day ahead. To serve, add ice cubes to four glasses, and pour over the Sling. Garnish with a citrus wedge. Add a (paper) straw. Enjoy.

SUMMER BEER SNACK

Crudités with herbed crème fraîche

You can always rely on a dip to get the party started. Simple to put together, and moreish to eat – what's not to love? Buy whatever fresh vegetables you can lay your hands on and, even better if they are locally grown with the earth still clinging to them. Just wash off any dirt and scrape or peel, then cut into batons for easy eating. I love crème fraîche for this dish, but you can use Greek yoghurt instead.

For the dip
200 gm crème fraîche
3 spring onions, finely chopped
2 tbsp each coarsely chopped dill, mint and flat-leaf parsley leaves
2 tbsp lemon juice
Salt and pepper

To serve:
A selection of crudités, such as radish, carrot, cucumber, celery, young asparagus

Put the crème fraîche into a bowl. Stir in the onions and the herbs. Add the lemon juice and season. Arrange your raw vegetables around the dip and start scooping – but no double-dipping, mind!

Autumn

Autumn is for eating well and eating cleverly, with its nutrient-dense vegetables such as squash and kale, all the better for boosting your immune system. Think filling, warming and nutritious, perfect for when the nights start drawing in.

Saying goodbye to summer is always a tough one but if I had to pick my favourite time of the year for eating, then it's the autumn. The summer crops might be coming to an end, but the season is still full of abundance, while winter crops are underway as the days shorten. It's also the time to preserve for the leaner growing months ahead. Freezing, fermenting, and pickling for later consumption is itself a hugely rewarding activity.

WHAT'S IN SEASON?

Apples • Assorted herbs • Beetroot • Broccoli • Cabbage • Carrots • Cauliflower • Celeriac • Celery • Chard • Chestnuts • Chillies • Courgettes • Cranberries • Cucumber • Fennel • Figs • Garlic • Globe artichokes • Jerusalem artichokes • Kale • Kohlrabi • Leeks • Marrow • Medlar • Mushrooms • Onion • Pak choi • Parsnips • Pears • Peppers • Plums • Potatoes • Quince • Raspberries • Rocket • Runner beans • Salsify • Shallots • Spinach • Squash • Swede • Sweetcorn • Tomatoes • Turnip • Watercress

AUTUMN FORAGING

My top 5: sloes, rosehips, common hogweed seeds, rowanberries, hazelnuts

Clouds of snowy white flowers that cover **BLACKTHORN** trees might be one of the highlights of spring, but come the autumn, the tree is made even more appealing thanks to its rich, dark, inky fruit. I'm talking **SLOES**. Is there a better foraged cocktail than a sloe gin fizz (see page 170)? First you've got to pick your sloes and steep them in gin for a bit (see page 172) but it's easy, and the rewards are great. The sloe is an ancestor of the cultivated plum, and bears fruit in September and October, but it's best picked after the first frost as it makes the skins softer. Another fun fact – the tree has long been associated with witchcraft. Legend has it that witches' wands and staffs are made of blackthorn wood.

Did you know that **ROSEHIPS** contain 20 times more vitamin C than oranges? Rosehips are an unsung hero. You'll commonly find them in hedgerows and on the edges of woodland. Along canals and rivers, they're mostly found in dog rose plants. You can use them in several ways, to make tea, or in jams and jellies, plus there's rosehip syrup, of course, made popular during rationing in World War II as a replacement for hard-to-get-oranges. Little tip: make sure they are red in colour and soft to the touch before picking.

COMMON HOGWEED is another popular foraged food, growing prolifically in hedgerows and rough grassland. A member of the carrot family, it displays large, umbrella-like clusters of creamy-white flowers mostly between May and August, the plant growing up to two metres tall. It's not to be confused with highly poisonous giant hogweed, which grows up to five metres, and whose sap will burn your skin. That said, it's best to use gloves when handling common hogweed as some people can react to it. Why bother? For me, it's all about the fragrant cardamom-like seeds harvested from the flower heads in the autumn. Blend them up in a spice grinder and add them to cakes, hot chocolate and flapjacks.

I first came across **ROWANBERRIES** used in cooking at Fäviken. The legendary Swedish restaurant (since closed) run by charismatic chef Magnus Nilsson had a jar of them pickling in the kitchen. The Scandinavians love rowanberries, but us Brits, not so much. The fruit is bitter, even when cooked with plenty of sugar. And cook them you must, as they contain harmful toxins, which cooking denatures. But if you like bittersweet flavours as I do, you might want to give them a try. They're best for making jelly or schnapps, both great served with cheese. The rowan tree, or Mountain Ash, grows at high altitudes, in woodlands, and in gardens and parks. It's easy to spot from a distance with its clusters of red or bright orange berries come autumn.

If I had to pick, then the **HAZELNUT** would be my favourite nut. When I'm feeling flush, I'll buy a bag of Piedmont hazelnuts, arguably the world's best, but when I find them ripe in the wild in late September through to October, it's like I've struck gold. Also known as filberts and cobnuts, they grow as far north as Scotland, but crop more regularly further south, and are found in woodland areas, scrubland and hedgerows. The trick is to get the ripe ones before the squirrels and birds do – not that you want to deprive them of food, so always leave plenty for them. Though you can get a jump by picking them while still green and leaving to ripen in a warm dark place, turning often, before removing the hard outer shell. I love them best on my morning porridge chopped and lightly roasted (see page 100).

MAKE YOUR OWN FORAGED HAZELNUT BUTTER

First remove the hard outer shells to get to your nuts. Preheat the oven to 180°/fan 160°/gas 4. Spread the shelled nuts on a piece of baking paper on a baking tray and toast for eight minutes until lightly browned. Once cooled, tip them onto a tea towel and rub off the skins gently before transferring to a food processor. Blend for about 10 minutes until a butter forms. Add a pinch of salt, a pinch of sugar and sometimes I like to add a pinch of cinnamon, too. Blend again. Then decant into a clean jar or slather over toast immediately.

Strapatsada

BREAKFAST

I love this classic Greek breakfast dish. The word originates from the Italian *strapazzare*, meaning, in the context of eggs, to scramble, so I guess they enjoyed this dish too, at some stage in history. And it's not hard to see why – the combination of creamy eggs with piquant tomato and savoury feta is a winner that will set you up for the day. There are lots of different variations of the dish, many using fresh tomatoes cooked down first. But for boat cooking ease, I've simplified it rather, then pimped it up with a topping that will deliver a visual treat to match the gustatory one.

FOR 2 SERVINGS

200g cherry tomatoes, halved
1 tbsp extra virgin olive oil
Pinch of dried oregano
Salt and black pepper
6 eggs
2 garlic cloves
20g butter
1 tbsp tomato purée
50g feta
Handful of dill, chopped

Heat the grill to high. Put the tomatoes in a dish, cut-side up, then drizzle with oil, dried oregano, and season with salt and pepper. Grill for 7-8 minutes, or until charred and juicy. Crack the eggs into a bowl and whisk together. Finely slice the garlic. Melt the butter in a non-stick frying pan over a medium heat and add the garlic, cooking for a couple of minutes until fragrant. Add the tomato purée and cook for another minute, then reduce the heat and fold in the eggs, cooking gently until creamy. Take it off the heat, stir in the cherry tomatoes, and season with salt and pepper to taste, then plate up and garnish with crumbled feta, dill and another twist of black pepper.

Avocado and dukkah toast

BREAKFAST

I'm a big fan of dukkah, sprinkling it on everything (for the recipe, see page 21), but I like it best for breakfast with avocado. I sometimes add a poached egg or a soft-boiled egg too if I'm feeling hungry, or a smear of cream cheese underneath the avocado. Either way, it will add another dimension to your breakfast.

FOR 4 SERVINGS

2 ripe avocados
Juice of 1 lemon
½ tsp chilli flakes
Salt and pepper
4 large slices of sourdough bread
2 tbsp dukkah

Scoop the flesh out of the avocado and add to a bowl. Add the chilli flakes and lemon juice then mash everything together with the back of a fork, and season. Toast the bread and smooth over the mashed avocados and sprinkle with the dukkah.

Mushroom kimchi rarebit

QUICK & LIGHT

This is definitely not your traditional Welsh rarebit. It doesn't contain stout, nor does it have rarebit staples Worcestershire sauce and English mustard. Instead, kimchi is the star flavour kick here. When kimchi is combined with melted cheese, a delicious alchemy happens. Remember that the next time you make a toasted cheese sandwich.

FOR 4 SERVINGS

250g mature cheddar, grated
3 large eggs, lightly beaten

1 egg yolk
125g kimchi, chopped
8 large flat mushrooms
Green salad to serve

Preheat the oven to 200°/fan 180°/gas 6. In a bowl, combine the grated cheese, eggs, egg yolk, and kimchi. Pull out the mushroom stems (keeping the trimmings for another dish) and place in a tray lined with baking paper. Divide the filling between the mushrooms and cook for 20-25 minutes, until puffed up and golden. Serve with a green salad.

Tomato tapenade tart

QUICK & LIGHT

I always keep a jar of tapenade in the cupboard. It's great for impromptu suppers, or when you want a savoury kick to any sandwich. You will find this olive spread all over southern Europe, but best is the classic Provençal black olive tapenade. The combination of black olives, capers, garlic and herbs, and often anchovies, is addictive. You can make your own, but the bought ones are generally good and a worthy addition to your store cupboard – especially when combined with roasted tomatoes and crisp flaky pastry. Use whatever tomatoes you can lay your hands on and serve with a green salad.

FOR 4 SERVINGS

320g pre-rolled puff pastry sheet
170g jar of black tapenade
3 medium tomatoes, sliced
6 cherry tomatoes, halved
Maldon salt and black pepper
1 tbsp extra virgin olive oil, plus extra for serving
Handful of basil leaves, torn

Preheat the oven to 220°/fan 200°/gas 7. Line a 38 x 25 cm baking tin with baking parchment. Place the pastry on the parchment and score a 2 cm border all around the edges of the pastry. Spoon a generous layer of tapenade inside the scored border then arrange the tomatoes over the top in a single layer, using the smaller tomatoes to fill in any gaps. Season with salt and pepper, and drizzle with oil. Bake the tart for 25 minutes, or until the pastry borders are puffed and crisp, the base is a light golden brown, and the tomatoes are soft and charring at the edges. Remove the tart from the oven, and scatter over the torn basil. Allow it to cool a little in the tray before slicing into rectangles. Drizzle with a little more oil to serve.

Greek spinach rice and grilled lamb chops

BIGGER PLATES

Spinach rice, or spanakorizo, is one of my favourite things to eat when I visit Greece. You need a medium grain rice, such as Italian Arborio. In Greece, it's often served with slow-cooked lamb. I serve it with crumbled feta on top, or like this, with grilled lamb chops. Some farmers get their lamb ready for Easter, but the best time of the year to eat British lamb is autumn, after they've been bouncing around lush green fields all summer. Ask the butcher for best end cutlets with the bones trimmed of fat and skin (to avoid any flare-ups) and let the meat rest for 10 minutes before scoffing.

FOR 4 SERVINGS

450g spinach, rinsed
2 lemons
1 medium onion, chopped
2 tbsp extra virgin olive oil
1 tsp dried mint
2 tbsp chopped dill
120g risotto rice
Salt and pepper
4 best end lamb cutlets
Tomato salad to serve (optional)

In a large pot, wilt the spinach with the juice from half a lemon and a teaspoon of olive oil. Set aside to drain. In the same pot, sauté the onion with the rest of the olive oil until soft. Add the spinach, mint and dill to 320 ml warm water and bring to the boil. Add the rice, salt to taste, pepper and simmer, covered, for about 20 minutes until the rice is soft. Add additional warm water if needed. Serve warm with a squeeze of lemon and a drizzle of oil.

Meanwhile, cook your chops. Preheat the barbecue, griddle pan or grill. Season both sides of the chops. Place on the grill or griddle pan, and brown on each side for about three minutes, squeeze over a little lemon juice while grilling. Serve with the spinach rice, and a tomato salad too if you like.

Chef Recipe – Sat Bains

Minced venison and oyster mushrooms

BIGGER PLATES

Two Michelin-starred chef Sat Bains is used to seeing boats gliding by his eponymous restaurant with rooms on the River Trent, even welcoming a fair few boaters to eat. The towpath runs along the bottom of his impressive kitchen garden. Sat is a fit guy, weight training since he was 14 years old, and always at the gym when he's not in the kitchen. So it came as a huge shock when he had a heart attack at the age of 50. He made a full recovery, thank goodness, and while doing so wrote a cookbook, *Eat to Your Heart's Content* (published by Kyle Books), which aims to improve heart health without compromising on taste. It more than succeeds in its mission. This recipe, which he has let me share, is one of my favourites. Only a chef of his calibre could come up with this particular combination of punchy ingredients. It's simple, too, and calls for just one pan. I've ditched the 30g of salted butter stirred into the mushrooms, but your call.

FOR 2 SERVINGS

30 ml extra virgin olive oil
400g venison mince
200g oyster mushrooms, thinly sliced
Maldon sea salt and black pepper
1 tsp ground cumin
1 tsp smoked paprika
1 tsp sesame oil
1 tsp fish sauce
1 tsp rice wine vinegar
4 spring onions, thinly sliced
1 tbsp crème fraîche (or I sometimes use yoghurt) to serve
Juice of 1 lime

For the wet mix:
2 large shallots, thinly sliced
25g minced garlic
20g minced ginger

2 bird's eye chillies, finely sliced (or 1 regular chilli if you prefer less heat)
Iceberg lettuce or flatbreads, crème fraîche or natural yoghurt, and a squeeze of lime juice to serve

Heat the olive oil in a non-stick frying pan over a medium-high heat, then add the venison mince and fry until golden and separated. Add the mushrooms and cook for 10-12 minutes, until the mushrooms are browned. Season with salt and pepper, and add the cumin and smoked paprika, cooking for a further four minutes. Now add the wet mix and cook for around eight minutes or until the vegetables are soft. Add the sesame oil, fish sauce and rice wine vinegar, and cook for another two minutes. Turn off the heat and allow the mince to cool slightly. Add the chopped spring onions. Serve scooped into Iceberg lettuce cups or with flatbread, plus a dollop of crème fraîche (or yoghurt) and a squeeze of lime.

Basque cheesecake

SWEET TREATS

With its caramelised top and jiggly centre, the Basque cheesecake has captivated the food world in recent years, with each kitchen carefully guarding their own recipe. Basically, it's a heat-blasted custardy pud that contains just four ingredients – soft cheese, eggs, cream and sugar. This recipe throws tradition to the wind with its inclusion of boat-friendly tinned sweetened condensed milk, but it tastes pretty good and has just three ingredients. You can use the leftover condensed milk in your porridge for breakfast the next day. I sought further advice from the three Basque sisters at La Maritxu bakery in London, who specialise in this dessert – theirs is outstanding. So I refined the recipe by lessening the cooking time, increasing the amount of cream cheese in the mix (I use Philadelphia Original), and making sure the ingredients are used at room temperature.

FOR 6 SERVINGS

2 x 280g full fat soft cheese, at room temperature
3 large free-range eggs, at room temperature
175 ml sweetened condensed milk

Preheat the oven to 220°/fan 200°/gas 7 and line a 20cm springform cake tin with one piece of baking paper (to stop any leaks) – it doesn't matter if it's a bit creased, it's a rough and ready dessert. Put all the ingredients into a large bowl and whisk together until thoroughly combined (at home, use a food processor). Pour the batter into the prepared cake tin and bake for 15-18 minutes or until the cake is dark brown, puffy around the edges, and still jiggly in the middle. Let it cool completely in the tin before eating (it's supposed to deflate, so don't worry). For the best flavour, refrigerate for a few hours to let the flavours combine.

Honey-roasted figs

SWEET TREATS

Is there a more beautiful fruit than a fig? With its soft, sweet reddish flesh and crunchy seeds, it tastes and looks richly exotic. And because of climate change, it now grows well in the UK's warmer climes. The plant loves a sunny wall best, trained as a fan, its roots restricted so it concentrates on the fruit. In the UK, it crops twice a year, but outdoors only once, mine ripening in the late summer and early autumn. I love figs cut into wedges and served with air-dried ham, such as Spain's highly prized Pata Negra. I add figs to salads and eat them for breakfast with a spoonful of Greek yoghurt, sprinkled with chopped toasted nuts. But my favourite way to eat figs is baked just like these until lightly caramelised and drizzled with honey, and served as an accompaniment to cheese.

FOR 4 SERVINGS

8 fresh figs, halved lengthways
2 tbsp unsalted butter
2 tbsp honey
Maldon salt
Cheese to serve

Preheat the oven to 180°/fan 160°/gas 4. Place the figs cut-side up on a baking tray lined with baking paper. Dot the butter over evenly and drizzle over half the honey. Give each fig half a little sprinkle of salt. Bake for about 20 minutes until they are soft and slightly caramelised. Remove from the oven, drizzle over the rest of the honey and let them cool to room temperature. Serve with cheese.

Plum and coconut cake

SWEET TREATS

I've adapted this recipe from one given to me by the breakfast chef at the Hotel Rathaus Wein & Design in Vienna (the Austrians make the best cakes). Their version had drained pitted cherries on it and it was topped with a chocolate ganache – yes, for breakfast! It was so good. I use the basic batter all the time for a quick fruit-topped cake, which I sometimes finish with a drizzle of melted dark chocolate if there is something to celebrate. I've tried it with blueberries (both frozen and fresh), apricots, fresh cherries (and tinned cherries), blackberries, both wild and cultivated – and now plums. It always works. It's ridiculously easy and quick to pull together, using just one bowl, and at around 30 minutes in the oven it's a good one for the boat. I've added mahlab spice to this, ground up cherry stones that have an aromatic almond taste and are used a lot in Greek and Middle Eastern baking, but you can use vanilla essence instead.

FOR 6 TO 9 SERVINGS

100g full fat yoghurt
100g vegetable oil
100g milk
100g desiccated coconut

1 tsp ground mahlab
150g caster sugar
200g plain flour
1 tsp bicarbonate of soda
300g stoned, quartered plums

Preheat the oven to 200°/fan 180°/Gas 6. Combine all the ingredients except the plums in a mixing bowl and pour the batter into a lined 23cm loose-bottomed baking tin. Arrange the plums on top and bake for about 30 minutes, or until the cake is lightly browned and the plums start catching. Let it cool in the tin for a few minutes before transferring to a wire tray to cool completely.

TIPPLE TIME

WHAT TO DRINK

As the temperature drops, so do the leaves, along with apples and pears, and a whole host of other exciting autumnal ingredients. I let the seasonal ingredients dictate what I'm going to drink.

By the first cold snap, I'm already after more weight in my wines. For mushroom risotto, say, I'm looking at earthier Pinot Noirs to start my early autumn drinking, along with Barbera from Northern Italy, with its lively acidity and berry, cherry notes. And a special shout out for autumn favourite, Grenache, which goes all plush, spicy and opulent when it jumps into bed with Syrah and Mourvèdre, demanding rich, meaty stews. By then, I'm moving towards big reds from sunnier climates, such as Shiraz, Cabernet Sauvignon and Merlot. You get the idea.

The autumn is not just for red wines, it's full-bodied white wine season, too. Say hello to Sémillon, particularly those from California and Pessac-Léognan in Bordeaux. And meet Marsanne and Roussanne, a couple of Northern Rhône whites that shine when blended and make a perfect partner for roast pork with a side of squash. Don't overlook Portuguese white wines, either. Europe's most underrated wine region offers everything from zesty complex whites from Bairrada, to the rich, full-bodied whites of the Douro and Dão.

As the nights get chillier, a beer shift is also in order. Cue brown ales, which offer more robust flavours, a little sweetness on the palate and a dry, crisp finish – a great match for autumn suppers. And I do love a dark beer – I'm talking porters, stouts and black beers, which add another layer of flavour. Porter and stouts have a reputation as the desserts of the beer world, and they do indeed pair well with puds – chocolate fondant particularly – but a creamy stout with a plate of oysters is also a joy; while a dark beer with plate of smoky spareribs is a match made in heaven.

And nothing says autumn like cider, as the fruit continues to be picked through to October. There are many different styles around now, but there are foods to fit them all, from mushroom dishes to creamy chicken, roast pork belly, and pot roast pheasant. Cider is also a dream with bonfire night favourites, squash and sausages. Seek out locally produced ciders – the proper farmhouse stuff, which is undergoing a renaissance (see page 60).

Autumn is also the time to infuse spirits with your hedgerow finds, from damson and sloe gins to wild blackberry liqueur and elderberry wine. All you need is a fermenting bucket, a demijohn, and 2kg of elderberries. Cheers!

THREE GREAT AUTUMN COCKTAILS

Kir royale

This vividly coloured cocktail has long been associated with the Burgundy town of Dijon. The story goes that it was named after Canon Félix Kir, a Catholic priest and hero of the French resistance during World War II. The original version is crème de cassis (blackcurrant liqueur) topped with Champagne. My version is made with English sparkling wine and a British-made blackcurrant liqueur from White Heron in Herefordshire. With its single, perfect wild blackberry as a garnish, it's the classiest autumn cocktail.

25 ml White Heron British Cassis, or to taste
English sparkling wine or Champagne to top up, chilled
1 blackberry to garnish

Pour the blackcurrant liqueur into a chilled flute, then top up with the sparkling wine and garnish with a single blackberry.

Sloe gin fizz

Another vibrantly coloured drink – this time hot pink. The sloe gin fizz is one of the most British of cocktails. Made with sloe gin, which is now widely available to buy, or better yet, make it yourself (see page 172), it ticks a lot of boxes.

50 ml sloe gin
25 ml freshly squeezed lemon juice
2 tsp sugar syrup
1 egg white
Soda or sparkling water
Slice of lemon to garnish

Add the gin, lemon juice, sugar syrup and egg white into a cocktail shaker. Shake without ice first to emulsify the egg white with the liquids, then shake

again with ice to chill the drink. Strain the contents of the shaker into the glass. Top with the soda or sparkling water. Garnish with a slice of lemon.

Mulled apple juice

All the fun of the season, but without the booze. But if you would rather have a boozy one, then replace the apple juice with cider.

FOR 4 GLASSES

1 litre of apple juice
1 orange, zest pared into strips
1 star anise
3 cloves
2 cinnamon sticks, snapped in half
Honey to taste

Simmer the apple juice with the orange zest, star anise, cloves and cinnamon sticks for about five minutes. Add honey to taste and serve in mugs, each with a little peel and cinnamon.

HOW TO MAKE SLOE GIN

Technically a fruit liqueur, sloe gin is easy to make yourself – you just need to wait until the sloes are ripe enough and then wait again for the sloe gin to macerate, which takes about two or 3 months.

Between the end of September and November, wild blackthorn trees are heavy with sloes. Traditionally picked after the first frost, which breaks down the fruit's thick skin, they grow all around the country. When ripe, sloes are a rich, dark purple colour and are soft to the touch like a plum. Eaten raw, they are tart and bitter, but once steeped in gin the warmth of the fruit comes through, along with some stunning ruby to pink hues. With an abv of between 15% to 30%, sloe gin has hints of wild cherry, pomegranate and cranberry on the palate.

Most recipes add sugar to the gin to begin maceration, but I prefer to add it at the end to let the natural fruit sugars do their thing. Sweetening the sloe gin at the end means you can control the sweetness levels better. This is one to make at home rather than on board as you will need a freezer for this. You will also need a 2 litre wide-mouthed glass jar.

As well as using sloe gin to make an array of cocktails (a Sloe Royale is another favourite, just add Champagne), including the famous sloe gin fizz (see page 170), you can also add sloe gin to a Tom Collins, Gimlet and even a Negroni. Or try making a sloe liqueur – raid your drinks cabinet for any old bottles of brandy or whisky, then follow the same method. Or just sip it neat.

300g ripe sloes
70 cl bottle of good quality gin
Simple syrup (made with around 200g caster sugar, see below) and/or honey to taste

Freeze the sloes to mimic the first frost for a minimum of 24 hours (48 hours is best). Place the frozen sloes into a wide-mouth lidded jar, filling it about halfway. Pour in a good quality gin to fill to the top. Leave in a cool, dark place for two to three months, turning the jar a couple of times. Line a sieve with a square of muslin set over a bowl and strain the sloe gin. Sweeten to taste with the simple syrup, or honey, or both. It's now ready to drink.

PERFECT SIMPLE SYRUP

Combine equal measures of sugar and water in a saucepan over a low heat. Warm the mixture until the sugar dissolves, then allow it to cool. Transfer to a glass container. Store in the fridge for 2-3 weeks.

AUTUMN BEER SNACK

Black bean and corn quesadillas

As a beer snack, quesadillas rule. They hail from the northern regions of Mexico and are popular in Tex-Mex cuisine. Quesadilla translates as 'little cheesy', as it's primarily filled with cheese, but also sometimes meat, beans, vegetables and spices. I love this combination of cheesy beans, sweetcorn and Mexican spices. I use a shop-bought fajita spice mix for this, but you could make up your own.

FOR 4 SERVINGS

400g tin of black beans, rinsed and drained
260g tin of sweetcorn, drained
3 spring onions, finely sliced
Handful of coriander, washed and chopped (optional)
250g mature cheddar, grated
2 tsp fajita seasoning
Salt and pepper
8 wholemeal flour tortillas

In a bowl combine the beans, sweetcorn, spring onions, coriander if using, cheddar and spice mix, then season. Divide the mixture between four of the tortillas leaving a small gap around the edge. Top with the remaining tortillas. Heat a non-stick frying pan over a medium heat, slide in the first quesadilla and cook on each side, flipping carefully with a spatula, until golden brown and crisp. Slice each quesadilla into triangles and serve straight away, nibbling as you cook the remaining quesadillas.

Winter

Embrace the winter season with nourishing foods that satisfy appetites. The cold weather makes stomachs grumble more than in the summer months. And no, carbs are not your enemy, they're a smart survival strategy.

Hello, comfort food. I'm talking slow-cooked stews, melty cheese and steaming potatoes, hearty soups et al. This doesn't mean nutritionally challenged stodge – there are plenty of ways to make comfort food healthy, from a warming bowl of porridge served with poached fruit, to a bolstering bowl of beans. There's also a whole array of cold season veg that comes packed with the good stuff, from high-fibre cauliflower and beetroot to vitamin-packed spinach and kale. Plus, there's no limit to the number of ways these vegetables (and it is mostly veg at this time of the year) can be prepared and enjoyed, including roasting, baking, slow-cooking, and airfrying. As the weather gets colder and the days shorten, what better way to spend your time than in the kitchen or galley, cooking your favourite winter warmers.

> **WHAT'S IN SEASON?**
>
> Beetroot • Brussel sprouts • Carrots • Cauliflower • Celeriac • Celery • Chestnuts • Jerusalem artichokes • Kale • Kohlrabi • Mushrooms • Onions • Parsnips • Potatoes • Salsify • Savoy cabbage • Shallots • Swedes • Swiss chard • Turnips • Winter squash

WINTER FORAGING

My top 5: acorns, wintercress, hawthorn berries, chestnuts, three-cornered leeks

ACORNS were once consumed in abundance back in the Middle Ages as they're packed with nutrients, such as iron and manganese. The pigs used to make prized Spanish Iberico ham feast on acorns. Korean culture prizes the acorn and it's a staple in the cuisine. I'm not talking raw acorns, which are unsafe to eat due to their tannin levels, but cooked acorns, which frankly are a bit of a faff, involving soaking, steeping, and boiling before blending – and only then are you ready to cook with them. Wild food enthusiasts use acorns for making pastry and bread. For acorn pastry, measure out two parts acorn mush to one part plain flour, before rubbing in butter. What does the pastry taste like? Dense, crumbly, nutty.

This is a useful one to know: **WINTERCRESS**. It's commonly grown, available all year round where the winters are mild, and typically found on riverbanks or indeed anywhere damp, with its rocket-shaped leaves, yellow flowers, and its distinctly hot, pleasantly bitter flavour that's not dissimilar to watercress. You can use it the same way too – add it raw to salads, or cook it up like spinach. A great find over winter when other greens have disappeared, it's one of several types of cress knocking about and was grown as a food until

the early 1900s. Its leaves have a hefty dose of vitamins A and C. Best though are the unopened flower buds, which pack a punch in the flavour department. Apparently, the ancient Greeks used wintercress as an aphrodisiac – just saying.

If you like sweet and sour flavours, you'll love **HAWTHORN** berries, or haws, as foragers call them. They're the tiny red or orange fruits that grow on trees and shrubs that belong to the crataegus genus, the hawthorn trees. The fruit ripens through to September but can stay on the tree until late winter. Tasted raw, the haws are unpalatably sour and can upset your stomach, but when cooked they offer a sweet and sour, spicy flavour not unlike rosehips. Try making ketchups with them, or purée and use as a base for pizza, or slathering on burgers, even flavouring brandy. Hawthorn berries have been used in medicine for centuries to help lower blood pressure and anxiety, and improve circulation, and they are bursting with vitamin C and rich in polyphenols. Little tip: to check for optimum ripeness, squeeze the fruit gently until it reveals the soft custard yellow flesh underneath.

CHESTNUTS smell like Christmas to me, especially when sellers roast them over braziers on cold December evenings as the tree lights twinkle. And thanks to my increasingly gluten-intolerant mates, I use chestnuts fairly frequently in dishes – boiled, puréed, grilled, steamed, deep-fried, candied, and ground into flour. Excellent in ice cream and stuffing, and positively dreamy in a cake, chestnuts are extremely versatile – nutritious, too, as they are a great source of potassium, plus other minerals and vitamins, and low in fat. Chestnuts were an integral part of Roman cuisine, which they brought with them to Britain. Be sure to score the skin first so they don't explode on the open fire, or cook in a heavy frying pan instead.

The **THREE-CORNERED LEEK** is the latest foraged food to grace tasting menus. It's the allium every top chef wants – and everybody can get, thanks to its prolific growth from November to early spring, on roadsides, parks, fields and near pathways. It's easy to identify, too – it looks like a white bluebell but has a green stripe running through the white petals, and a garlicky smell when you crush the leaf in your hand. All the plant is edible. The flowers are great in salads, the leaves in soups and stews, while the more mature onion-like root can be used like spring onions or garlic. It's also great pickled, fermented, or cooked on a barbecue, while the flowers make a beautiful garnish. But I love it best when freshly chopped and stirred through mashed potatoes.

Tofu, kimchi and kale scramble

BREAKFAST

Tofu, or bean curd, is made by curdling fresh soya milk then pressing it into a block and cooking it – not unlike the way cheese is made. Used extensively in Korean, Thai and Chinese cooking, it's a great source of protein and rich in nutrients. Its texture, too, is wonderfully adaptable, from smooth and creamy, to crisp and crunchy. I mostly eat it like this, as a scramble – it's quick and easy and can be pimped up with numerous healthy additions, from spinach and cherry tomatoes, which I add in the summer months, to this kale and kimchi combo in the winter. Serve for breakfast with a slice of toasted sourdough.

FOR 4 SERVINGS

1 tablespoon sesame oil
400g firm tofu, drained, crumbled between your fingers
1 tbsp tamari soy sauce
½ teaspoon garlic powder
3 big handfuls of chopped kale, large stems removed, washed
3 spring onions, trimmed and sliced
3 tbsp kimchi, roughly chopped
Handful of coriander, finely chopped, to serve (optional)

In a frying pan, heat the sesame oil over a medium-high heat. Add the tofu and cook, stirring occasionally, for about 2-3 minutes. Turn the heat down to medium, add soy sauce and garlic powder, and give it a good stir. Add kale and the spring onions. Cook, stirring occasionally, until the kale becomes soft and wilted – you can cover partially to help with the process, about 5-7 minutes. Lastly, add the kimchi, mix and cook just until the kimchi has warmed through, about a minute or so (you don't want to overcook the kimchi, or it will destroy some of its health benefits). Serve with the chopped coriander, if using.

Chickpeas, chorizo and eggs

BREAKFAST

I love a bit of spice for breakfast, and chorizo delivers just the right amount. Though you can add a pinch of chilli flakes to spice things up further if you fancy. It's basically a twist on a good old British fry up, or North African egg dish shakshuka – take your pick, but it's one that works for lunch or supper as well as brekkie. You can't beat a good Spanish chorizo but if you want to buy British then I love the Suffolk Salami Co Chorizo, which is widely available. I sometimes serve this with a pile of cooked greens, dressed with a little extra virgin olive oil and a squeeze of lemon.

FOR 4 SERVINGS

2 tbsp olive or rapeseed oil
1 onion, finely chopped
1 garlic clove, crushed
100g chorizo cut into 2cm pieces
2 x 400g cans chickpeas, drained and rinsed
125 ml tomato passata
Handful of parsley leaves, chopped
Salt and pepper
4 large free-range eggs

Heat oil in a large, heavy-based frying pan over medium heat. Add onion, garlic and chorizo, and cook, stirring, for 5-7 minutes until the onion begins to soften. Add the chickpeas and cook, stirring to coat, for a further 1-2 minutes. Add the passata and 250 ml water, then reduce heat to low. Cook for 10 minutes, or until slightly reduced. Add the chopped parsley, season and stir. Make four wells in the mixture and crack an egg into each. Cook until the whites are cooked and the yolk is still runny – if you add a lid it will speed up the cooking time to around 3-4 minutes. Serve immediately.

Leek and butter bean soup

QUICK & LIGHT

Tinned beans are my most used ingredient, in whatever form. And this soup is one of my favourites, inspired by a recipe from the late, great Italian food writer Marcella Hazan. I love the way the leeks collapse and intensify after just 15 minutes of gentle cooking. Then all you need is some stock and the drained white beans (any tinned white bean works here, even chickpeas). The result is one deeply flavoured, immensely satisfying soup. You can add a handful of grated Parmesan if you have some, but it doesn't need it. Mop the lot up with some crusty bread.

FOR 4 SERVINGS

3 medium leeks, sliced finely and washed
1 tbsp olive oil

Pinch of salt and black pepper
1 litre vegetable stock
2 x 400g butter beans, drained and rinsed (I love Bold Bean Co. Queen Butter Beans)
1 handful of flat leaf parsley, chopped

Soften the leeks in the oil over a low heat, adding a pinch of salt, with the lid on, for 15 minutes. Add the stock and the drained beans. Heat through for five minutes. Add a grind of black pepper, stir through the chopped parsley and serve.

Fi B's fishy rice

QUICK & LIGHT

My old mate Fiona Beckett (matchingfoodandwine.com) stands by this budget store cupboard lunch. It's perfect for boat cooking, especially when you haven't got much in the way of anything fresh. You can pimp it up or dress it down, and tins of flavoured fish work best here, such as in tomato sauce, or lemon. Use freshly cooked rice for this recipe, rather than leftover rice. And keep a tub of supermarket crispy onions on the go to add texture and flavour. If you want to ramp things up further, stir through a tablespoon of chopped kimchi.

FOR 2 SERVINGS

150g basmati rice
2 x 120g tins of flavoured sardines
Salt
1 lemon or lime (optional)
1 small handful of chopped coriander, or parsley
1 heaped tbsp crispy, fried onions

Cook the rice according to the instructions on the packet and drain, keeping it warm. Tip the tinned fish including some of its oil into the rice and mix thoroughly, breaking up the fish as you go. Season with a little salt and a good squeeze of lime or lemon. Add the chopped herbs and stir through the warm rice. To finish, sprinkle over the crispy onions.

Broccoli and anchovy toasts

QUICK & LIGHT

I'm a big fan of piling stuff on toast. It works for any occasion – breakfast, brunch, lunch, supper, and especially pre-dinner nibbles. Not a day goes by when I'm not eating something loaded onto toasted sourdough. And it is usually sourdough. But if I can't lay my hands on an artisanal loaf then toasted supermarket ciabatta works just fine. From homemade baked beans spiked with chilli, to roasted tomatoes dressed in crème fraîche and chopped mint – it all goes on toast. I particularly love scrambled eggs in a miso-laden brown butter, and rarebit any which way, using leftover cheese scraps and adding softened leeks or kimchi for a bit of a kick. Avocado and tahini-roasted chickpeas is also a regular toast-topping. For something more devilish, try melted dark chocolate with a drizzle of olive oil and sea salt – you can thank me later.

Broccoli and anchovy are the perfect partners. Bung them on bread and it's a satisfying snack, one I often eat pre-dinner, cut into small squares, Prosecco in hand. And we have the Italians to thank in part for this one, who use the combination widely on pasta and in cicchetti (savoury Venetian bar snacks). You need to get over your fear of cooking the broccoli longer than you normally would as you want it soft enough to mash with the back of a fork. Ideally serve while still warm, or at room temperature. To drink? Great with a crisp, dry white wine.

FOR 4 SERVINGS

350g small broccoli florets
Flaky sea salt and freshly ground black pepper
Pinch of chilli flakes, or to taste
2 small (28g) tins of anchovy fillets in olive oil, drained and chopped
1 ciabatta, sliced and toasted
½ lemon
6 basil leaves, torn (optional), to serve

Bring a pan of salted water to the boil and add the broccoli florets. Simmer for six minutes, or until the florets are fork soft. Drain thoroughly. Turn into a bowl and mash with a fork, crumble in the sea salt, add a pinch of chilli flakes, the chopped anchovies and black pepper, then pile onto the ciabatta toasts and squeeze over the lemon. Serve with the torn basil leaves, if using.

Chef Recipe – Michel Roux

Truffade

BIGGER PLATES

Michel Roux is a figurehead of British haute cuisine. So what's he doing in my humble little book, you might ask? Well, he took to the water in a narrowboat on the River Thames to spread the love and share some great recipes in a television show for The Food Network called *Roux Down the River* (available to watch on Amazon Prime) after closing his family's iconic Mayfair restaurant Le Gavroche. "I do enjoy being on the water. You find you take a breath, and life goes at a different pace. It's a nice time to relax and reflect," says Roux. I couldn't agree more.

Along the way he found some great spots to eat, and, at the epicentre, the Roux family's three Michelin-starred Waterside Inn, arguably the country's poshest riverside dining address.

Roux also drew inspiration from the local produce that he found, using it in the recipes that he cooks back on board. These 'cheesy chips', as he prefers

to call his truffade, hail from the Auvergne and are one of his favourites, which he kindly allowed me to use for the book.

The potatoes are cooked in lard or oil with bacon and garlic, and then topped with cheese, ideally a mix of Red Leicester for colour and bite, Emmental, and finally Ogleshield for stringiness, the West Country's answer to raclette. Heart stoppingly good.

FOR 4 SERVINGS

1.2kg Maris Piper potatoes (or roasting potato equivalent)
4 tbsp lard or oil
50g bacon lardons (optional)
Salt and pepper
400g cheese, such as Emmental, Comté, Spenwood, Ogleshield or Cheddar, grated
2 garlic cloves, minced
Green salad and cold cuts to serve (optional)

Wash and peel the potatoes, then cut into thin slices about 3-4mm. Add the lard or oil and the lardons (if using) to a cast iron casserole and cook on a medium heat until the lard has melted. Add the potatoes, season with salt and pepper to taste, and give it all a stir. Cover and cook on a medium to low heat for 25 minutes, stirring gently from time to time, or until cooked through. Heat the grill, then add the cheese and garlic to the potatoes, giving it all another good stir, and brown under the grill for a couple of minutes. Serve immediately, ideally with a crisp, vinegary, green salad and some cold cuts.

Chef Recipe – Judy Joo

Korean beef stew

BIGGER PLATES

The universal popularity of Korean barbecue means the country is celebrated for its mastery of beef. And it's not all about fieriness in Korean cuisine, as this popular umami-rich stew shows. Otherwise known as kalbi jjim, the depth of flavour from the soy-enriched stock is addictive.

This recipe comes from chef Judy Joo, a Korean American Londoner, and Wall Street analyst turned chef and restaurateur, who is a regular on TV screens both in the UK and the US. Our friendship was cemented on a trip together to South Korea for an assignment some years ago.

I've talked about how useful a pressure cooker is on a boat (see page 18) – this would normally take three hours in the oven, but it takes less than half an hour using this method.

Ribeye trimmed with the bone in is the most authentic cut, but use whatever diced beef you can lay your hands on. Just don't be tempted to skip the browning bit as the caramelisation adds another layer of flavour.

It doesn't really need rice, but I often serve it with jasmine rice or basmati. If you like, add a simple cucumber salad dressed with a little soy, rice vinegar, sugar and a pinch of chilli powder. And garnish the stew with sesame seeds and spring onions for that Korean-style flourish.

FOR 4 SERVINGS

1 tsp Maldon salt
800g diced beef
2 tbsp roasted sesame oil
1 small onion, diced
3 tsp garlic purée (or 2 garlic cloves, crushed)
100 ml soy sauce
1 tsp ground ginger (or 2 tsp fresh ginger, grated)
50g brown sugar
1 tsp freshly ground black pepper
300 ml beef stock (I use Knorr)
1 large carrot, peeled and cut into 2cm pieces
500g of baby potatoes
Sesame seeds (optional)
Spring onions, sliced (optional)

Crush the salt with your fingers and sprinkle over the beef. Heat the sesame oil in the pressure cooker then brown the beef. Remove from the heat. Add the onions, garlic, soy sauce, ginger, brown sugar, black pepper and stock, giving it all a good stir. Add the carrots and potatoes. Place it back on the hob with the lid, sealing it shut, and bring it up to pressure. Set the timer for 13 minutes, then turn off the heat and allow it to cook for a further 10-15 minutes as the pressure naturally releases. Now it's ready to serve.

Blood orange and poppy seed muffins

SWEET TREATS

Blood oranges are in season from December to May, and at their sweetest and most plentiful in January and February when the temperature difference between day and night is largest. They look like regular oranges from the outside, but inside the flesh can be blood red, though despite the name they come in a variety of shades, from barely blushing to deep ruby red. If you can't find any for this quick, easy muffin recipe, a normal orange works just fine. Enjoy for breakfast or afternoon tea, or for dessert with a dollop of Greek yoghurt.

MAKES 12 MUFFINS

300g self-raising flour
150g golden caster sugar
1 heaped tsp baking powder
3 tbsp poppy seeds
5 tbsp sunflower oil
1 large egg
200 ml milk
2 blood oranges, zested and juiced
125g granulated sugar

Preheat oven to 200°/fan 180°/gas 6. Mix the flour, caster sugar, baking powder and poppy seeds together in a large bowl. In another bowl, whisk the oil, egg, milk, zest and juice of one of the oranges. Pour the wet ingredients into the dry and combine until it just comes together. Spoon into muffin papers and bake for 12-15 minutes, until risen and golden, then transfer to a wire rack. While the muffins are baking, in a small bowl mix the remaining zest and juice with the granulated sugar. After the muffins have cooled for a couple of minutes, spoon the juice and sugar mix over each muffin. Leave to set for five minutes before scoffing.

Chocolate, pistachio and apricot tiffin

SWEET TREATS

Tiffin, or fridge cake, is a British tea-time classic. While the word tiffin originates from India (an Anglo-Indian word for a light meal or snack), chocolate tiffin is a different beast. Traditionally, it's a no-bake fridge cake that as legend would have it was first created by a bakery in Scotland in the 1900s. My version is a tad healthier, replacing the biscuit with oats – and yes, there is a little baking involved.

I usually make this at home and take it on board. It stores happily in an airtight container somewhere cool, and keeps me going for a few days. But if you have a fridge on board, then you're good to go, as it needs to set in the fridge first before slicing and storing for future nibbling. You can play around with the nuts and dried fruit, but this is my preferred combo.

MAKES 16 SLICES

100g jumbo oats
100g pistachios (or walnuts, or almonds), roughly chopped
50g mixed seeds
2 tbsp coconut oil
150g dark chocolate (at least 70% cocoa solids), broken into chunks
100g dried apricots (or prunes), chopped

Preheat the oven to 200°/fan 180°/gas 6. Line a baking tin or dish, about 20 x 20cm, with baking parchment. Dry roast the oats, nuts and seeds on a large baking sheet for about eight minutes, until the oats are crispy, and the nuts are lightly browned. Tip onto a plate and leave to cool. Melt the coconut oil and chocolate together in a heatproof bowl over a saucepan of gently simmering water. Remove from the heat and stir in the toasted oats, nuts, and seeds. Add the chopped apricots. Spread the mixture evenly in the prepared tin and leave to set in the fridge. Cut into squares when completely cold and store in an airtight container.

TIPPLE TIME

WHAT TO DRINK

When the wind is up and the temperature drops, it's time to hunker down on board with brown spirits and warm spices. Think hot buttered rum with a cinnamon stick, whisky mac with a crystallised ginger garnish, and brandy swirling in a glass.

Even gin can feel cosy and wintery with the right treatment – the botanicals come to life when heated, the juniper becoming suddenly warming. It's also time for Irish coffee and boozy hot chocolate, and vodka combined with the first of the season's blood oranges for a wintry take on a Collins.

And come the festive season, it's also the time for Champagne-based cocktails. One of the oldest, simplest, and best is a sugar-cube doused in aromatic bitters dropped into the base of a glass, a small measure of Cognac poured over it, before topping up with Champagne.

It's also the only time of the year I sip spirits neat, from Armagnac to

Cognac, Calvados and English apple brandy, rum and whisky. Add a crackling fire, a hooting owl, a few flickering stars, and it's winter-drinking nirvana.

Wine drinkers, too, can ramp things up on the texture front during the winter months. It's when full-bodied reds come out to play, along with buttery, oaky whites. Where to look? Check out the spicy, hefty reds from the Southern Rhône, which work particularly well with warming dishes of mushrooms and game. And look to the more ripe, generous Garnacha from Spain, which works a treat with grilled meat. Not forgetting Nebbiolo from the Piedmont region in Northern Italy, which is always a winner with winter's richer plates of food, such as sausages and lentils. Winter is also the time for a deep dive into big, jammy Malbec from Argentina, especially when you've got a stew on the go.

But if you're not wanting reds to drink with food, then steer towards wines that are a tad lighter on the tannins – look for wines aged in stainless steel, or more fashionable concrete. Whites can work just as well in winter if you seek out wines that are richer and more fruit forward. Think lightly oaked, buttery Chardonnay from Australia and California, South Africa and Burgundy. Chenin Blanc, too, offers more heft and substance at this time of the year, so as well as looking at those from the Loire, check out South Africa, where the fruit becomes more tropical in flavour. Full-bodied southern Italian whites, such as Grillo, also work well in the chillier months.

Beer has long been brewed for consumption over the winter months – my current favourite winter ale is Hook Norton's Twelve Days. Your winter ale is typically stronger, darker and sweeter, often with intense smoky, malty, leathery, and coffee notes – give it a try.

Soft options need be no less exciting, from mulled drinks made with pomegranate juice and all the usual spices, to fruit punch laced with ginger, shrubs and kombuchas.

THREE GREAT WINTER COCKTAILS

Hot gin skin

The hot toddy started life, or so the story goes, in British-controlled India in the 17th century. Derived from the Hindi word 'taddy', it was a drink made with fermented palm sap. By the 1820s the toddy had evolved into a mixture of alcohol, sugar, ginger and lime, but it wasn't hot. Perhaps the Scots were the first to heat it up? Who knows. The classic hot toddy that we know and love today is generally made with whisky. I love it best made with gin.

1 spoonful of honey
2 shots London dry gin
2 cloves
½ measure of lemon juice
1 lemon slice
1 cinnamon stick

Add the honey into a mug, along with the gin, cloves and lemon juice. Top with boiling water and stir until the honey is dissolved. Garnish with a lemon slice and a cinnamon stick.

Milk and honey

The classic version is made with just three ingredients – whisky, milk and honey, warmed gently together to make the ultimate nightcap or dessert replacement. You could make it like that, or you could skip a step and just use two ingredients, namely milk and Bénédictine, a famous French liqueur made from 27 herbs and spices and finished with honey. The elegant simplicity of it will win you over.

2 shots Bénédictine liqueur
Milk to top
1 orange slice and a cinnamon stick (optional) to garnish

Gently heat together Bénédictine and milk. Pour into a mug and garnish, if using, with the orange slice and a cinnamon stick.

Earl Grey and marmalade fizz

From January through to February, Seville oranges arrive from southern Spain, ready to be peeled, shredded and turned into marmalade – so why not add a little to a cocktail? I love the smokiness of the tea combined with the bright, citrussy flavours. The spicy warmth on the finish from the fresh ginger garnish elevates it further. **Little tip:** there are many different cocktails and mocktails that use tea as a base.

2-3 tbsp Seville orange marmalade
1 tbsp lemon juice
150 ml Earl Grey tea, cooled
A slice or three of fresh ginger to garnish

Put the marmalade, lemon juice and Earl Grey tea into a cocktail shaker with ice. Shake vigorously and strain into a large tumbler filled with ice. Garnish with slices of fresh ginger.

WINTER BEER SNACK

Montgomery Cheddar and rosemary popcorn

Named after the Cheddar Gorge caves in Somerset, where it traditionally used to be stored to ripen, Cheddar is the best-selling cheese in Britain. It accounts for half of all the country's cheese sales – and the entire top shelf of my fridge, thanks to my husband's obsession with toasted cheese sandwiches. Unlike the smooth, creamy texture of more mass-produced Cheddar, Montgomery's Cheddar has a dry, almost brittle feel, with a rich, savoury flavour. Made and perfected by three generations at Montgomery's 500-year-old farm in Cadbury, Somerset, it's one of the last three remaining traditional unpasteurised artisan Somerset Cheddars. And yes, it's off-the-scale good. But if you can't find it, just use whatever extra mature Cheddar you can lay your hands on.

FOR 4 SERVINGS

2 tbsp sunflower oil
250g popping corn
Pinch of Maldon salt
2 rosemary springs, pulled into smaller sprigs
25g unsalted butter, cut into small dice
100g Montgomery cheese, finely grated
Pinch of cayenne pepper

Heat the oil in a large saucepan over a medium-high heat. Add the popping corn, salt and rosemary and cover the pan with a lid. Once it starts popping, cook for 3-5 minutes, shaking the pan every 20 seconds. Remove the pan from the heat. Transfer to a large serving bowl, letting the unpopped kernels fall to the bottom. Add the butter to the same pan to melt, then pour over the popcorn adding the cheese, and cayenne pepper. Serve straight away.

BRITISH CHEESE

Did you know that there are now over 750 cheeses produced in the UK? We've come a long way since the Second World War, when its rationing system forced all cheesemakers to produce a single type, 'Government Cheddar'. Now British cheese is almost as thrilling as French. As well as revived traditional varieties, such as Cheshire and Red Leicester, there are a plethora of modern cheeses such as Cornish Yarg and Stichelton, which are sold in farm shops, delis, cheese counters and supermarkets up and down the country – so do seek out local cheese as you explore the country via its inland waterways. If I had to pick a top five, then these are it: Appleby's Cheshire, Tunworth, Berkswell, Montgomery's Cheddar and Isle of Wight Blue.

5 great places to buy British cheese (online or in person)

- Neal's Yard Dairy (London)
- Paxton & Whitfield (London)
- The Fine Cheese Company (Bath)
- I.J.Mellis (Scotland)
- The Courtyard Dairy (Lancashire)

ACKNOWLEDGEMENTS

If I hadn't visited *Ellenene* that chilly October weekend in 2021, then none of this would have happened. So, my biggest thanks go to my brother and sister-in-law, Kim and Liz Baxter, firstly for buying the boat, then for inviting me on board for that first weekend (and many subsequent ones) and for suggesting I write the book. Spending quality time with you both has been an absolute joy, but most importantly it gave me a deeper understanding and appreciation of our inland waterways.

Thanks also go to Hoseasons for arranging a stunner of a river boat on the Norfolk Broads for the 'renting a boat' section – talk about luxury living. *Evita* had all the mod cons and then some. The Norfolk Broads was at its magical best, and a must visit for any river boater.

But this book would not have been possible without the encouragement from the team at Bloomsbury/Adlard Coles, who saw its potential and let me run with it. It has been a blast. Great working with you.

A heartfelt thanks to legendary chefs Sat Bains and Michel Roux for sharing their winning recipes, ditto stellar chefs Lorcan Spiteri and Vanessa Marx, who have chosen to make our inland waterways part of their lives. Not forgetting chef Judy Joo, who passed on her obsession for kimchi and all things Korean – long may our travels together continue.

It was photographer Gary Latham who made most of these recipes come to life on the page with his keen eye for colour, composition and light, while Isle of Wight-based artist and illustrator Lilly Louise Allen has perfectly captured river life in her vibrant watercolours.

A third of the pictures were snapped by yours truly on my iPhone (easy to spot in comparison), although I hope that at the very least, they will show my enthusiasm for the recipes, which were enhanced significantly by the various props lent to me by dear mates Anna Hugo and Clare Spragge.

Finally, and most importantly, thanks to my lovely husband Mark, who encouraged me throughout and was a willing guinea pig for the dishes.

INDEX

A
Alphonso mango with coconut and mascarpone 111–12
asparagus and new potato frittata 105–7

B
banana pancakes with wild blackberries 130–2
barbecues 88–95
    cleaning 92
    fish 92
    meat 91
    sausages 91
    vegetables 93
Basque cheesecake 164–5
beef stew, Korean 190–3
blood orange and poppy seed muffins 194–5
bread-based recipes
    avocado and dukkah toast 156
    black bean and corn quesadilla 174
    broccoli and anchovy toasts 187–8
    naan bread breakfast pizza 101–3
    oaty soda bread 20–1
    panzanella 134–6
    sea salt brioche toasties 112
breakfast recipes
    avocado and dukkah toast 156
    banana pancakes with wild blackberries 130–2
    chickpeas, chorizo and eggs 182–3
    granola 129–30
    naan bread breakfast pizza 101–3
    porridge with rhubarb and orange compote and toasted almonds 100–1
    strapatsada 154
    tofu, kimchi and kale scramble 181
broccoli and anchovy toasts 187–8

C
chalk stream trout with fennel, herbs and citrus, roasted 136–8
cheese 204
chicken, spinach and lemon orzotto 107–8
chickpeas, chorizo and eggs 182–3
chocolate, pistachio and apricot tiffin 195–6
chutneys 51–2
cordials 56–7

D
dessert and sweet treat recipes 29
    Alphonso mango with coconut and mascarpone 111–12
    Basque cheesecake 164–5
    blood orange and poppy seed muffins 194–5
    chocolate, pistachio and apricot tiffin 195–6
    honey-roasted figs 166–7
    peaches with honey mascarpone and rosemary, grilled 140–2
    peanut butter biscuits 112–13
    plum and coconut cake 167–8
    scones, strawberries and clotted cream 139–40
    sea salt chocolate brioche toasties 112
dried herbs 54–6
drinks 59–63, 116–18, 143–4, 169, 197–9
    blood orange and vermouth cobbler 118–19
    citrus sling 147
    cordials 56–7
    craft beer 59–60
    craft cider 60–1
    Earl Grey and marmalade fizz 200–1
    elderflower cordial 57
    Ellenene fizz 144–6
    fruit liqueurs 58
    hot gin skin 199–200
    Kir royale 170
    milk and honey 200
    mulled apple juice 172
    peach Bellini 146–7
    rhubarb and lemon fizz 119
    rooibos and honey mocktail 121
    simple syrup, rhubarb 121
    sloe gin 172–3
    sloe gin fizz 170–2
    soft drinks 63
    spirits 62
    wine 61–2
dukkah 21–2
    toast 156

E
essential ingredients 20–25
equipment 16–19

F

ferments 54
figs, honey-roasted 166–7
fish recipes *see* seafood recipes
fishing in rivers 37–8
fishy rice 185–7
flowers 43
foraging 34–6
    autumn 152–3
    spring 98–9
    summer 126–8
    winter 178–9
fruit liqueurs 58

G

granola 129–30
growing food 40–4

H

herbs, dried 54–6
heritage, narrowboat 80–6
honey 45–6

I

ingredients
    essential 20–5
    growing 40–4
    local 30–3
    tinned 25–6

J

jam 48–51

L

leek and butter bean soup 183–5
light bite recipes
    asparagus and new potato frittata 105–7
    broccoli and anchovy toasts 187–8
    fishy rice 185–7
    hot smoked salmon salad with watercress and fennel 132–3
    leek and butter bean soup 183–5
    mushroom kimchi rarebit 156–7
    panzanella 134–6
    puy lentils with roasted sweet peppers, tomatoes and feta 133–4
    tomato tapenade tart 159
    wild garlic pesto 103–5
liqueurs, fruit 58

M

meat-based recipes
    chicken, spinach and lemon orzotto 107–8
    chickpeas, chorizo and eggs 182–3
    Greek spinach rice and grilled lamb chops 160
    Korean beef stew 190–3
    minced venison and oyster mushrooms 163–4
    rosemary, garlic and honey pork chops 138–9
mushroom kimchi rarebit 156–7

N

naan bread breakfast pizza 101–3
narrowboat heritage 80–6

P

panzanella 134–6
peaches with honey mascarpone and rosemary, grilled 140–2
peanut butter biscuits 112–13
pickling 48–58
    cucumbers 52–3
    plums 53
picnics 94–5
plum and coconut cake 167–8
porridge with rhubarb and orange compote and toasted almonds 100–1
prawn and chickpea stew with leeks and lemon 108–11
preserving 48–58
puy lentils with roasted sweet peppers, tomatoes and feta 133–4

Q

quick recipes
    asparagus and new potato frittata 105–7
    broccoli and anchovy toasts 187–8
    fishy rice 185–7
    hot smoked salmon salad with watercress and fennel 132–3
    leek and butter bean soup 183–5

mushroom kimchi rarebit 156–7
       panzanella 134–6
       puy lentils with roasted sweet peppers,
          tomatoes and feta 133–4
       tomato tapenade tart 159
       wild garlic pesto 103–5

R
renting boats 74–9
rosemary, garlic and honey pork chops
   138–9

S
scones, strawberries and clotted cream
   139–40
seafood recipes
       broccoli and anchovy toasts 187–8
       Fi B's fishy rice 185–7
       hot smoked salmon salad with
          watercress and fennel 132–3
       prawn and chickpea stew with leeks
          and lemon 108–11
       roasted chalk stream trout with fennel,
          herbs and citrus 136–8
sea salt chocolate brioche toasties 112
seasonal eating 33
       autumn 152
       spring 98
       summer 126
       winter 178
smoked salmon salad with watercress and
   fennel 132–3
snack recipes
       black bean and corn quesadillas 174
       crudités with herbed crème fraiche 148
       Montgomery Cheddar and rosemary
          popcorn 202
       prosciutto, crisps, Grindillas peppers
          122–3
soda bread, oaty 20–1
spinach rice and grilled lamb chops, Greek
   160
strapatsada 154
sustainable boating 70–2

sweet treat and dessert recipes 29
       Alphonso mango with coconut and
          mascarpone 111–12
       Basque cheesecake 164–5
       blood orange and poppy seed muffins
          194–5
       chocolate, pistachio and apricot tiffin
          195–6
       honey-roasted figs 166–7
       peaches with honey mascarpone and
          rosemary, grilled 140–2
       peanut butter biscuits 112–13
       plum and coconut cake 167–8
       scones, strawberries and clotted cream
          139–40
       sea salt chocolate brioche toasties 112

T
tofu, kimchi and kale scramble 181
tomato tapenade tart 159
truffade 188–90

V
vegetarian recipes
       asparagus and new potato frittata
          105–7
       black bean and corn quesadillas 174
       leek and butter bean soup 183–5
       mushroom kimchi rarebit 156–7
       panzanella 134–6
       puy lentils with roasted sweet peppers,
          tomatoes and feta 133–4
       strapatsada 154
       tofu, kimchi and kale scramble 181
       tomato tapenade tart 159
       wild garlic pesto 103–5
venison and oyster mushrooms, minced
   163–4
verjuice, crab apple 128
vinaigrette 23

W
wild garlic pesto 103–5